August 29, 2003

Dear Customer:

Thank you for your purchase of *Modelling Metabolism with Mathematica*, by Peter J. Mulquiney and Philip W. Kuchel.

The following includes several corrections to errors in the text. We sincerely regret any inconvenience this may have caused you. Please let us know if we can be of any assistance regarding this title or any other titles that CRC Press publishes.

Best regards,

CRC Press, LLC

Preface – Reference 6: Krebs, H.A.
Authors – P.W.K., line 6 should read: His post-doctoral work was at ...
Page 4 – Eqn {1.4} should read: $k_1/k_{-1} = \ldots$
Page 10 – *Mathematica* box, bottom line should read: ... a replacement rule
Page 13 – Down 6 lines of text: Insert 'k,' into the Clear function argument list.
Page 14 – Line 6 up: Insert 'k,' into the Clear function argument list.
A general principle to apply when implementing the Appendices is to 'Quit Kernel' and then Evaluate the relevant Appendix Notebook. This stores the compiled version in the 'correct' (determined by *Mathematica*) location on the hard disk.
Another general rule is to 'Quit Kernel' before starting a new chapter of the book, or in general do this if there are problems with evaluating an example.
Page 29 – Figure 1.1: Caption should read ... the Improved Euler method...; the right-hand arrowhead in the figure should point to the intersection of the vertical line and the line denoted L.
Page 31 – Line 3 up should read: ... information at (t_m/y_m) ...
Page 34 – Line 10 up should read: ... the automatic procedure ...
Page 35 – Line 4 up should read: ...circumstances ...
Page 37 – Reference 2 should read: *Schaum's Outline of Theories and Problems of Biochemistry*
Page 58 – Line 6 up: Insert 'k,' into the Clear function argument list.
Page 71 – Line 11 up: Insert 'k,' into the Clear function argument list.

#1468/0-8493-1468-2

Page 103 – Text box, line 7 should read: ... stoichiometry ...

Page 106 – Line 4 up should read: ... S_1 and S_4 to be ...

Page 113 – Lines 11 and 12 should read: ... since perturbation ... and ... the perturbation ...

Page 116 – Line 12 should read: Schuster

Page 122 – Figure 4.6 caption, line 1 should read: A_1 declines from ...

Page 130 – Section 4.9 heading should read: Decomposition ...

Page 133 – Line 17 up should read: Kacser

Page 138 – Line 2: The N-overbar matrix causes a problem in some computers so it is safest to replace it with the conventional *Mathematica* List form, namely, 'N-overbar = {{1, −1, 0}, {0, 1, −1}};' (the occurrence of this matrix format does not seem to cause any difficulties on page 146, however).

Page 153 – Reference 1: Kacser; Reference 7: Hofmeyr

Page 164 – Lines 15, 17, 18 should read: Eqn [6.20]

Page 170 – In the generation of 'noisy' data, change the 0.03 to 0.02; likewise on page 172 on two occasions. While not an error per se, this change makes the demonstration work cleanly on a repeatable basis. The variation in output is due to the random numbers that are generated.

Page 170 – Line 9 up should read: Using the function relevant to Eqn [6.24] ...

Page 265 – Line 3 should read: ... found in References 1–4 of Section 7.8; Thorburn and Kuchel (1985) *Eur. J. Biochem.* 150, 371–386; and McIntyre et al. (1989) *Eur. J. Biochem.* 180, 399–420.

Page 265 – Line 5: End paragraph after '7.1.'

Modelling Metabolism with *Mathematica*

Detailed Examples Including Erythrocyte Metabolism

Modelling Metabolism with *Mathematica*

Detailed Examples Including Erythrocyte Metabolism

Peter J. Mulquiney
Philip W. Kuchel

CRC Press
Taylor & Francis Group
Boca Raton London New York

CRC Press is an imprint of the
Taylor & Francis Group, an **informa** business

CRC Press
Taylor & Francis Group
6000 Broken Sound Parkway NW, Suite 300
Boca Raton, FL 33487-2742

First issued in paperback 2019

ISBN-13: 978-0-8493-1468-1 (hbk)
ISBN-13: 978-0-367-39522-3 (pbk)

Library of Congress Card Number 2003046279

Library of Congress Cataloging-in-Publication Data

Mulquiney, Peter J.
    Modeling metabolism with Mathematica / Peter J. Mulquiney, Philip W. Kuchel.
        p. cm.
    Includes bibliographical references and index.
    ISBN 0-8493-1468-2 (alk. paper)
        1. Cell metabolism—Computer simulation. 2. Enzyme kinetics—Computer simulation.
    3. Erythrocytes—Computer simulation. 4. Mathematica (Computer program language)
    I. Kuchel, Philip W. II. Title.

QH634.5.M85 2003
572'.4'0113--dc21   2003046279

**Visit the Taylor & Francis Web site at
http://www.taylorandfrancis.com**

**and the CRC Press Web site at
http://www.crcpress.com**

Preface

Simulation of Metabolism

The experimental and theoretical study of metabolism in mammalian cells, and the human erythrocyte in particular, has a long history, so it is valid to challenge the need for another book on this topic; but as you will see, this book is very different from previous ones in that it is interactive. Our response is that an understanding of cellular metabolism at the molecular level, with all its intricate controls, is far from complete and many fundamental and clinically relevant discoveries remain to be made. Three major technological advances in recent years – mass spectrometry, NMR spectroscopy, and computing – have greatly contributed to the renaissance of cellular metabolism as a topic of research. It is the computer modelling of metabolism, in particular the simulation of time courses of reactions, that is our focus.

Anticipated Readership

This book is aimed at advanced undergraduate and postgraduate students of biochemistry, enzymology, functional genomics, biotechnology, theoretical biology, computational science, applied mathematics – indeed, anyone interested in applying computational methods to the simulation and study of metabolic systems.

Contents

Chapters 1 and 2 provide an introduction to biochemical enzyme kinetics, including basic definitions; the mathematical formulation of reaction schemes and the computer-based methods used in their analysis; and quantitative aspects of enzymology such as the analysis of kinetic data, the mechanistic basis of enzyme inhibitions, and models of enzyme regulation. Chapter 3 contains an introduction to the procedures used to simulate metabolic systems using symbolic computation; and a model of the urea cycle of the human liver is used to exemplify these. More advanced methods incorporating matrix algebra are introduced in Chapter 4, where we show that the differential rate equations describing complex metabolic reaction schemes can be represented in a simple and compact way. Our implementation of the key elements of metabolic control analysis (MCA) as presented by Heinrich and Schuster[1] is described in Chapter 5. Parameter estimation is the subject of Chapter 6 and here we consider linear and nonlinear least-squares regression analysis for this purpose, and parameter estimation in large-scale metabolic networks where over-parameterization may be an issue. Chapter 7 applies the theory and methods developed in the previous chapters to the human erythrocyte to illustrate how a realistic model of metabolism can be built up. Finally, the concepts and methods described in Chapters 5 and 6 are used to perform MCA on the erythrocyte model presented in Chapter 7. The five appendices contain supplementary material and *Mathematica* code that is required to run the programs and worked examples contained in Chapters 1 – 8 from the interactive CD.

Layout

Each chapter is divided into sections for easy cross-referencing between topics. We have made extensive use of worked examples to emphasize, or to extend, a basic concept that is discussed in the body of the text. The exercises are posed as questions and are made to stand out from the main text by the use of a solid horizontal line and the letters Q (question) and A (answer) in the margin. This approach has been adapted from that used in our own[2] and other successful Schaum's Outline texts. In many cases the references that we cite are merely representative of the literature in a particular topic; we apologize in advance to those of our colleagues who feel their work has not been adequately referenced.

Motivation

Our motivation for writing this book was our success in modelling a particularly troublesome aspect of human erythrocyte metabolism, and our wish to share the methodology that led to the insights that we consider would have been unattainable by other means. Briefly, these insights were as follows.[3-5]

Sound explanations for several readily elicited metabolic responses in human erythrocytes had not previously been found, notably, the exquisite sensitivity of the steady-state concentration of 2,3-bisphosphoglycerate (2,3BPG) to pH, and to changes in oxygen partial pressure. Alteration of the intracellular pH from the normal value of 7.2 to 6.8 (only 0.4 pH units), which is a transition that is frequently encountered physiologically, brings about, over several hours, an almost total disappearance of 2,3BPG. Researchers sought an unidentified effector molecule that would be produced in a reaction that is under the control of the state of oxygenation of hemoglobin. It was surmised that this compound might decrease the activity of 2,3BPG synthase, or activate the 2,3BPG phosphatase that catalyses its hydrolysis, thus *linking* the oxygen partial pressure to the 2,3BPG concentration. On the other hand, we posited that an explanation of these two notable metabolic responses might be found by simply piecing together the vast and disparate metabolic and kinetic data available from almost a century of relevant scientific literature, and from our own experiments using modern analytical techniques. Hence, we combined our own NMR spectroscopy data of whole cells and cell extracts with computer modelling of the metabolism and provided a plausible explanation for the observations.[3]

NMR experiments of many types provide a means of rapidly, precisely, and in some circumstances uniquely, obtaining estimates of metabolite concentrations in a totally non-invasive way. As such, NMR methods admirably satisfy Krebs' notion of what constitutes the essential ingredient for scientific progress. Specifically, in summarizing his discovery of the urea cycle, Krebs wrote :[6]

"If there is a lesson to be drawn ... it is ... the importance to progress of new techniques, especially techniques which make it possible to conduct a large number of experiments, and of studying a phenomenon under many different conditions. ... It also illustrates the importance of following up an unexpected and puzzling observation arising in the course of the experiment. Luck, it is true, is necessary, but the more experiments are carried out, the greater is the probability of meeting with luck."

Another "method" that allows rapid evaluation of experimental data should be added to the list of techniques that Krebs did not use, namely, computer-based simulation of metabolic systems. We are not, of course, the first to suggest this approach. It was pioneered at the University of Philadelphia in the 1950s by Britton Chance and Joseph Higgins, and then in a very elaborate way by David and Lillian Garfinkel.[7] In the 1970s one of us (PWK) helped construct a computer model of the human urea cycle. However, a persistent criticism of this and all other computer models of metabolism has been the failure to address the consequences of intracellular partitioning of metabolites between the cytoplasm, mitochondria, and other organelles, and likely effects of the viscous intracellular milieu and surface adhesion, on the rates of enzymic reactions. Because it is non-invasive and highly selective to the detection of chemical species, NMR spectroscopy has provided a means to address many of these questions.

Also, in the present decade computers have gained so much in calculating speed and user friendliness that it is now routine practice, using a modest-cost personal computer, to calculate the time dependence of a kinetic system, whether physical, chemical, or biochemical, that is described by arrays of hundreds of stiff non-linear differential equations. In the past, the solutions were obtained using specialized programs, often written in machine code by the individual scientist. However, with the advent of sophisticated general programming environments like *Mathematica* that have implementations of contemporary algorithms and excellent graphics output capabilities, the task of developing new models of metabolism and visualizing their responses has become accessible to many students of biochemistry, and the life sciences in general.

Acknowledgments

This book has emerged from work carried out by past and present students in PWK's laboratory: accordingly, we gratefully acknowledge Michael York, Zoltan Endre, Glenn King, David Thorburn, Kiaran Kirk, Julia Raftos, Lisa McIntyre, Nicola Nygh, Serena Hyslop, Lindy Rae, and Hilary Berthon. They contributed kinetic data primarily from NMR spectroscopy and also assisted in the development of the computer model of human erythrocyte metabolism that culminated in PJM's Ph.D. thesis. We also thank Bill Bubb and Bob Chapman for their invaluable input into the NMR work, and Brian Bulliman and Bill Lowe for computing and technical assistance, respectively. David

Regan, who spent his Ph.D. with PWK studying diffusion in cellular systems using NMR and simulations, generously gave his time to proofread the entire book; and Professor Athel Cornish-Bowden of Marseille gave a thorough and incisive appraisal of an earlier draft and made numerous valuable suggestions for changes and inclusions. We thank Alan Henigman for his positive review of an early draft of this book and, with Julie Benner and Debra Pierce, assisted in the process of locating a suitable publisher. Fequierre Vilsaint of CRC Press took up the project with enthusiasm and we thank him, Pat Roberson, and Naomi Lynch enormously for their encouragement, support, and finally for delivering so efficiently the final product.

Mathematica

Finally, it is relevant to comment on *Mathematica* and how it relates to this book: we make extensive use of *Mathematica* in this book; in fact, it has been typeset in this program. Even in Chapter 1 there are examples of quite sophisticated usage of *Mathematica*. Therefore, it is recommended that the CD version of the book be read on your computer with *Mathematica* available to run the various examples and exercises.

As noted above, the main concepts of biochemical kinetics and metabolic control analysis are illustrated in the worked examples (marked with a Q) and are expanded upon in the exercises that are located at the conclusion of each chapter, all using *Mathematica* in some way. We encourage you to not only reproduce the output of these examples but to make modifications to the programs and explore the results of these changes. A spin-off of all the biochemical, metabolic, and kinetic investigations will be the development of a facility in the use of this truly outstanding program.

If, on occasion, a *Mathematica* function is encountered that requires further clarification, simply proceed to the Help Menu and look up the index. Alternatively, a hard copy of the *Mathematica* book can be used, but after some practice you will most likely find the screen version to be more convenient.

The easy-to-use text by Don[8] is a valuable source of descriptions of the various functions and semantics of *Mathematica* and for illustrating their usage. There are also many specialized texts, like the present one, that use *Mathematica* to present their particular scientific or mathematical themes and that can provide examples and tricks of usage that are seemingly infinite in number. We anticipate that it will become obvious to you that scientists have gained, in *Mathematica*, access to mathematical 'power' and resources that were undreamed of even 5 years ago.

Conclusions

Mathematica, like all computer languages, is best learnt by experimentation; likewise for gaining skills in metabolic simulation. Therefore, we encourage you to run as many of the examples in this book as possible, and to be creative in your exploration of these and with the exercises that are listed at the end of each chapter.

Finally, we will welcome feedback from you on how this book could be improved.

Peter J. Mulquiney
Philip W. Kuchel
Sydney 2003

References

1. Heinrich, R. and Schuster, S. (1996) *The Regulation of Cellular Systems*.Chapman & Hall, New York.
2. Kuchel, P.W. and Ralston, G.B. (Eds.) (1998) *Schaum's Outline of Theory and Problems of Biochemistry*. 2nd ed. McGraw-Hill, New York.
3. Mulquiney, P.J., Bubb, W.A., and Kuchel, P.W. (1999a) Model of 2,3-bisphosphoglycerate metabolism in the human erythrocyte based on detailed enzyme kinetic equations: *in vivo* kinetic characterisation of 2,3-bisphosphoglycerate synthase/phosphatase using ^{13}C and ^{31}P NMR. *Biochem. J.* **342**, 567-580.
4. Mulquiney, P.J. and Kuchel, P.W. (1999b) Model of 2,3-bisphosphoglycerate metabolism in the human erythrocyte based on detailed enzyme kinetic equations: computer simulation and metabolic control analysis. *Biochem. J.* **342**, 597-604.
5. Mulquiney, P.J. and Kuchel, P.W. (1999c) Model of 2,3-bisphosphoglycerate metabolism in the human erythrocyte based on detailed enzyme kinetic equations: equations and parameter refinement. *Biochem. J.* **342**, 581-596.
6. Kebs, H.A. (1976) The discovery of the ornithine cycle. In: *The Urea Cycle*, Grisolia, S., Baguena, R., and Mayor, F., Eds. John Wiley & Sons, New York, pp. 1-12.
7. Garfinkel, D., Garfinkel, L., Pring, M., Green, S.B., and Chance, B. (1970) Computer applications to biochemical kinetics. *Ann. Rev. Biochem.* **39**, 473-498.
8. Don, E. (2001) *Schaum's Outline of Theory and Problems of Mathematica*. McGraw-Hill, New York.

Authors

Peter J. Mulquiney, B.Sc. (Hons) Ph.D., is a C.J. Martin Fellow of the Australian NH&MRC in the School of Molecular and Microbial Biosciences at the University of Sydney. He undertook his Ph.D. in biochemistry with Philip Kuchel in the area of metabolic simulation and NMR spectroscopy of human erythrocyte metabolism. A major outcome of this work was the first quantitive model of the regulation of 2,3-bisphosphoglycerate in the human erythrocyte. His post-doctoral work was at the University of Oxford where he worked with Kieren Clarke on NMR spectroscopy of the mammalian heart and computer modelling of cardiac metabolism.

Philip W. Kuchel, B.Med.Sc. (Hons) M.B., B.S., Ph.D., F.A.A., is McCaughey Professor of Biochemistry, and from 2003-2007 an ARC Australian Professorial Fellow at the University of Sydney, where he has been since 1980. He obtained his biochemical, medical, and early mathematical training at the University of Adelaide, and his Ph.D. in physical biochemistry at the Australian National University in 1975 with Laurie Nichol and Peter Jeffrey. His post-doctoral was at the University of Oxford, where he began his NMR studies of cellular systems with Frank Brown, Iain Campbell, and Dallas Rabenstein. Major discoveries and developments in his group have included the 'split peak phenomenon,' diffusion of solutes in cells, diffusion coherence in erythrocyte suspensions, and computer models of metabolism based on NMR data. He is a former president of the Australian Society for Biophysics and the Australian Society for Biochemistry and Molecular Biology.

Table of Contents

Chapter 1 - Introduction to Chemical Kinetics and Numerical Integration

1.1 Aims and Objectives		1
1.2 Complexity		2
1.3 Definitions		3
	1.3.1 Principle of mass action	3
	1.3.2 Equilibrium constant	4
	1.3.3 Molecularity	5
	1.3.4 Order of a reaction	5
	1.3.5 Units of rate constants	6
	1.3.6 Extent of reaction	6
1.4 Time Courses of Reactions		7
	1.4.1 Introduction	7
	1.4.2 Half-life	10
	1.4.3 Reaction lifetime	11
	1.4.4 Coupled first-order reactions	12
	1.4.5 Multiple-coupled reactions	16
	1.4.6 Non-first-order systems	18
1.5 Numerical Integration of Differential Equations		20
	1.5.1 General	20
	1.5.2 Numerical integration - overview	22
	1.5.3 Taylor series solution	25
	1.5.4 Runge-Kutta methods - overview	26
	1.5.5 Improved Euler method	28
	1.5.6 Modified Euler method	29
	1.5.7 General Runge-Kutta methods	30
	1.5.8 Higher-order Runge-Kutta methods	31
1.6 Predictor Corrector Methods		31
	1.6.1 The predictor	32
	1.6.2 The corrector	33
	1.6.3 Choosing the value of h	34
	1.6.4 Numerical integration in *Mathematica* - **NDSolve** and stiffness	35
1.7 Conclusions		35
1.8 Exercises		36
1.9 References		37

Chapter 2 - Elements of Enzyme Kinetics

2.1 Kinetics of Enzymic Reactions		39
	2.1.1 Purpose	39
	2.1.2 The Michaelis-Menten equation	39

2.1.3   Graphical evaluation of V_{max} and K_m 41
2.1.4   Lineweaver-Burk plot   41
2.1.5   Eadie-Hofstee plot   42
2.1.6   Hanes-Woolf plot   44
2.1.7   Eisenthal and Cornish-Bowden equation (direct linear plot)   44
2.2 Enzyme Inhibition   44
2.2.1   Degree of inhibition   44
2.2.2   Michaelis-Menten equations that include inhibitor effects   45
2.3 Enzyme Mechanisms   45
2.3.1   Michaelis-Menten mechanism   45
2.3.2   The steady state   47
2.3.3   Reversible Michaelis-Menten enzyme   48
2.4 Regulatory Enzymes   50
2.5 Exercises   51
2.6 References   51

Chapter 3 - Basic Procedures for Simulating Metabolic Systems

3.1 Introduction   53
3.1.1   Inborn errors of metabolism   54
3.1.2   Constancy of K_{eq} 54
3.2 Relationships between Unitary Rate Constants and Steady-State Parameters   55
3.2.1   Progress Curve of a Michaelis-Menten reaction   55
3.2.2   Pre-steady-state Michaelis-Menten scheme   57
3.2.3   Enzyme oligomerization and the turnover number   59
3.2.4   Specific examples of enzyme mechanisms   59
3.3 Upper Limit of Values for Unitary Rate Constants   63
3.3.1   Diffusion control of reaction rate   63
3.4 Realistic Enzyme Models   65
3.4.1   Deriving steady-state rate equations   65
3.4.2   The **RateEquation** function   67
3.4.3   Calculating a consistent set of unitary rate constants   69
3.5 Deriving Expressions for Steady-State Parameters   69
3.6 Multiple Equilibria   75
3.7 pH Effects on Kinetic Parameters   79
3.7.1   Ionization of the substrate   79
3.8 A Simple Model of the Urea Cycle   81
3.9 Conclusions   89
3.10 Exercises   90
3.11 References   93

Chapter 4 - Advanced Simulation of Metabolic Pathways

4.1 Introduction   95
4.2 Simulating the Time-Dependent Behavior of Multienzyme Systems   95
4.3 Using Matrix Notation in Simulating Metabolic Pathways   97
4.4 Generating the Stoichiometry Matrix   102

4.5 Determining Steady-State Concentrations 105
4.6 Conversation Relations 110
4.7 Stability of a Steady State 113
4.8 When Cell Volume Changes with Time 118
4.9 Decomposition of N and Calculation of the Link Matrix (Optional) 130
4.10 Exercises 132
4.11 References 132

Chapter 5 - Metabolic Control Analysis

5.1 Introduction 133
5.2 Control Coefficients 134
5.3 Calculation of Control Coefficients by Numerical Perturbation 140
5.4 Elasticity Coefficients 142
5.5 Response Coefficients 145
5.6 Internal Response Coefficients 149
5.7 Conclusions 151
5.8 Exercises 152
5.9 References 153

Chapter 6 - Parameter Estimation

6.1 Introduction 155
6.2 Approaches to Parameter Estimation 155
6.3 Least Squares 156
6.4 Maximum *a Posteriori* (MAP) 157
6.5 Parameters in Rate Equations 159
    6.5.1 Linear least squares 159
    6.5.2 Nonlinear estimation 162
    6.5.3 Nonlinear least squares 164
    6.5.4 Nonlinear MAP 166
6.6 Parameters in Systems of Differential Equations 167
6.7 Optimal Parameters 171
6.8 Variances of Parameters 172
6.9 Exercises 173
6.10 References 174

Chapter 7 - Model of Erythrocyte Metabolism

7.1 Introduction 175
7.2 Models of Erythrocyte Metabolism 175
7.3 Stoichiometry of Human Erythrocyte Metabolism 177
7.4 *In Vivo* Steady State of the Erythrocyte 182
7.5 Conservation of Mass Relationships 187
7.6 Simulating a Time Course 192
7.7 Exercises 195
7.8 References 196

Chapter 8 - Metabolic Control Analysis of Human Erythrocyte Metabolism

8.1 Introduction   197
8.2 Normal *In Vivo* Steady State   197
8.3 Identifying Zero Fluxes   198
8.4 Flux Control Coefficients   200
8.5 Concentration Control Coefficients   206
8.6 Response Coefficients and Partitioned Responses   207
8.7 Elasticity Coefficients   211
8.8 Internal Response Coefficients   212
8.9 Concluding Remarks   213
8.10 Exercises   214
8.11 References   215

Appendix 1 - Rate Equation Deriver

217

Appendix 2 - Metabolic Control Analysis Functions

223

Appendix 3 - Rate Equations for Enzymes of the Human Erythrocyte

265

Appendix 4 - Initial Conditions and External Parameters for the Erythrocyte Model

291

Appendix 5 - Equation List Describing the Erythrocyte Model of Chapters 7 and 8

295

Index

299

1 Introduction to Chemical Kinetics and Numerical Integration

1.1 Aims and Objectives

This book is about simulating the chemical dynamics of metabolic pathways using computer methods. It develops the many required concepts by using numerous examples and has as its ultimate goal a model of the metabolism of the most intensively studied mammalian cell, the human erythrocyte.

Our three main aims in writing the book were, first, to present a complete set of concepts required for the deterministic theory of chemical and enzyme kinetics, in order to equip students of Biochemistry to formulate their own dynamic models of time-dependent metabolic systems. An enabling objective was that a student would become proficient with *Mathematica* since it is in this 'environment' that the book is written. We chose *Mathematica* because of its huge suite of mathematical functions that can be implemented by simple one-line commands; for its symbolic computational power; and for the large number of other texts that can be used by a student to gain proficiency in the language and to glean ideas for metabolic modelling.

The second aim was to present in detail a realistic and contemporary computer model of metabolism, and for this we chose one on human erythrocytes. To our knowledge, this model[1] is one of the most comprehensive yet produced. We found that a model that includes the *fine details* of pH effects and Mg^{2+} binding to metabolites on the enzymes of the erythrocyte was necessary in order to provide reliable chemical interpretations of experimental data that were routinely obtained when NMR spectroscopy was applied to these cells. In other words, the model includes the pH dependence of various reactions, considers the concentration of Mg^{2+}-metabolite complexes, and it successfully predicted outcomes under unexplored situations that were subsequently verified experimentally.

Even if you, the reader, are not interested in erythrocyte metabolism *per se,* the act of developing the model, which is fully described, may yet be of interest, as the various processes involved are applicable to metabolic models of any pathway in any cell type. A second enabling objective was to encourage the understanding of the modern paradigm of metabolic analysis, namely, metabolic control theory or analysis (MCA). Accordingly, *Mathematica* is used to implement all the major procedures of MCA; this type of 'meta-analysis' is implemented and applied to the model of erythrocyte metabolism.

The third aim was a biochemical one: to summarize and document a consistent set of rate equations and associated kinetic parameters for most of the enzymes of the human erythrocyte. Inevitably, in the literature, there are instances of wildly different estimates of enzyme kinetic parameters and conflicting conclusions drawn from kinetic studies. Thus, much of the significance of the present model lies in the arguments surrounding our choice, for the model, of particular choices of values of the various kinetic parameters. By a process of iteration between data from a wide range of experimental situations and the simulations of these experiments, key kinetic parameters were adjusted in order to obtain a fit of the model to 'experimental reality.[1] The speed with which *Mathematica* can solve a large array of non-linear differential equations that make up the model was and is crucial to this iterative process. The present model can be simulated for 10 h of 'real' metabolism in ~1 min of computer time; so the effect of altering a kinetic-parameter value on metabolite concentrations can be assessed very rapidly. This quick response is vital when applying MCA, and it helps identify those parameters that are most responsible for a given metabolic response. This analysis also facilitates the choice of parameters to include in any simplified model derived for some special purpose from the more complex one; this is called *model reduction*.

1.2 Complexity

It is evident that in the operation of metabolic pathways, as in most complex systems, there are readily describable features or responses that are not inherent in the kinetic behavior of the individual enzymes, when they are studied in isolation. The responses of the system as a whole can manifest as unanticipated fluctuations in metabolite concentrations, including stable periodicities, growth in size of micro-, meso-, or even macroscopic structures, and possibly even self-replication of the system.

There are basically three patterns, or networks, of interconnection that operate within metabolic pathways in cells. The first is the interconnectedness that is represented by the chemical transformations that are predicated on the laws of chemical reactivity; the reactions involve substrates that become products, that in turn become substrates for other enzymes. The second entails reactants that activate or inhibit, and thus *control*, the enzymes of the pathway at places in the sequence that are often far removed, chemically, from the enzyme-catalyzed reactions that produce them. And the third level involves the relative abundance or amounts of individual enzymes that are determined by their rates of breakdown and rates of synthesis, via proteases and via mRNA, respectively. Preceding the translation of mRNA into protein is the regulation of transcription of DNA by transcription factors, many of which are hormone- and metabolite-binding proteins. Hormones and metabolites from outside a metabolic pathway therefore *regulate* it by bringing about changes in enzyme concentrations.

These three levels of interconnection are what shape the kinetic behavior of a metabolic pathway. The responses of the pathway to perturbations of enzyme activity, rates of substrate supply or product removal, do not succumb readily to intuitive analysis, at least not quantitative analysis; the systems are just too complex for that. The situation is akin to what Aristotle alluded to when considering certain features of geometrical systems: "The whole is greater than the sum of the parts." For metabolic pathways, the adage can be fruitfully reworded to: "The responses of the system are *different* from those predicted from the responses of the parts when they are studied alone." For example, individual enzymes in a closed system never display regular periodic variations in reactant concentrations, but sequences of enzymic reactions in a thermodynamically open system can.

Thus the philosophical stance taken in this book is the *reductionist* one, which in the present context amounts to the statement: "Notwithstanding the complexity of metabolic systems there are no 'hidden forces' that come into play when the system reaches a certain level of complexity." However, we acknowledge that an operational definition of complexity is not easy to formulate either!

1.3 Definitions

There are several key concepts that are essential for an understanding of chemical and enzyme kinetics; they are as follows.[1]

1.3.1 Principle of mass action

This is the fundamental principle of chemical and enzyme kinetics. It provides the procedure that is used to write down the mathematical equation(s) that describe the rate of a chemical reaction. It is "The rate of a chemical reaction is proportional to the product of the concentrations of the reactants involved in the elementary chemical process." The constant of proportionality is called the *rate constant,* or more specifically the *unitary rate constant* in order to emphasize the fact that it applies to an elementary process. A subtlety that may need consideration, especially when describing reactions in concentrated biological solutions, is that it is *chemical activity* and not simply concentration that should actually appear in the rate equations.

Chemical activity is given by the product of the concentration and a dimensionless factor called the *activity coefficient.* For most biological systems the activity coefficient is taken to be 1. However, even a simple solution such as 154 mM [or 0.9% (w/v)] NaCl, which is known as "physiological saline," has an experimentally measured osmolality of 283 mOsmol kg^{-1} at 25°C, compared with what would be obtained with the ideal solution of $2 \times 154 = 304$ mOsmol kg^{-1}. Thus the operational non-ideality coefficient for dilute solutions of this simple salt is ~0.93.

Q: What is the formalism used to represent a chemical reaction, and how do we express its time dependence?

A: Consider the reversible reaction scheme with reactants A, B, P, and Q, where a common convention is to denote substrates as capital letters from the first part of the alphabet and products by letters farther down the alphabet. The rate constants k_1 and k_{-1} *characterize* the rate of the forward and reverse reactions, respectively.

$$A + B \underset{k_{-1}}{\overset{k_1}{\rightleftharpoons}} P + Q \ , \tag{1.1}$$

$$\text{forward rate} = k_1[A][B] \ , \tag{1.2}$$

$$\text{reverse rate} = k_{-1}[P][Q] \ . \tag{1.3}$$

The square brackets denote concentration in mol L^{-1}.

At chemical equilibrium, the forward and reverse reaction rates are equal, so there is no *net* production of any of the reactants with time. Hence,

$$\frac{k_1}{k_{-1}} \frac{[P]_e[Q]_e}{[A]_e[B]_e} = K_e \ , \tag{1.4}$$

where the subscript e indicates that the concentration is evident at equilibrium, and K_e is called the *equilibrium constant*.

The reaction rate is usually expressed as the change in concentration of a chemical species per unit of time, so it is written mathematically as a *derivative*. The mathematical expression for the *net* rate of change of [A], therefore, includes terms to describe both forward flux (Greek: 'to flow'; units mol $L^{-1} s^{-1}$) *from* A, and reverse flux *to* A; e.g., for the above reaction the rate expression for [A] is

$$\frac{d[A]}{dt} = -k_1[A][B] + k_{-1}[P][Q] \ . \tag{1.5}$$

1.3.2 Equilibrium constant

The equilibrium constant of a chemical reaction, including an enzymic reaction, is defined as the product of the concentration of the products divided by those of the substrates, when the reaction is at equilibrium. A subtle question that is prompted by this definition, in relation to enzymes and macromolecules that bind ligands such as hemoglobin that binds oxygen, is: Which species should be defined as the product(s) and which the substrate(s)?

In a reaction between enzyme E and reactant A that can be represented as $E + A \rightleftharpoons EA$, we might expect to consider the product to be EA; however, in enzyme kinetics it is conventional to consider the products to be free E and free A and to view the EA complex as the *primary reactant*. Thus the reaction is described in terms of the *dissociation* of the EA complex; hence, the equilibrium constant is called a *dissociation equilibrium* constant. Such constants have the same units as concentration, namely, mol L^{-1}.

1.3.3 Molecularity

This appellation, when applied to a specific reaction, refers to the number of molecules involved in the *elementary* reaction. In a simple decomposition process that involves only one molecule, called *fission* or *scission,* the molecularity is one. However, most reactions such as hydrolytic ones involve two molecules colliding so the molecularity is 2. Another example is that the molecularities of the forward and reverse reactions depicted in Eqn [1.1] are both 2.

1.3.4 Order of a reaction

The *overall* order of a reaction is equal to the sum of the powers to which the concentrations are raised in the rate equation. The order with respect to individual reactants is equal to the powers to which their concentrations alone are raised.

Q: What is the overall order of the reaction that involves the condensation of A with B and whose rate is described by the following equation?

$$\frac{d[A]}{dt} = -k_1[A]^{1/3}[B]^{1/2} \ . \tag{1.6}$$

A: 5/6.

The overall order of the reaction is the sum of the exponents of the two concentration terms, namely, $1/3 + 1/2 = 5/6$. The order of the reaction with respect to A is 1/3, and for B it is 1/2.

It is usual for a simple reaction to have a molecularity that is an integer, but it is important to note that the order of a reaction is an experimentally determined value. The simplest way to determine it is to measure the reaction rate for a range of reactant concentrations and then to plot log(rate) versus log(concentration) and this gives a line with a slope equal to the order. If the concentration of all the substrates are altered in a constant ratio then the *overall reaction order* is determined. If just one concentration is varied, while the rest are held constant, then the reaction order for just that species emerges from the analysis.

Q: How do we experimentally determine the order of a bimolecular reaction?

A: Suppose that the two reactants are A and B. The order with respect to A is determined by fixing the concentration of B and measuring the initial rate of decline of [A] from its initial concentration $[A]_0$; and vice versa when determining the order of reaction for B. A graph of log(− rate of decline) versus $\log([A]_0)$ has a slope equal to the order. This can be seen from the general expression for the rate which is a generalization of Eqn [1.6].

$$\text{rate} = \frac{d[A]}{dt} = -k_1[A]^n[B]^m \quad,$$

$$\log(-\text{rate}) = \log(k_1) + n\,\log([A]_0) + m\,log([A]_0) \quad, \tag{1.7}$$

$$= n\,log([A]_0) + \text{Constant} \quad.$$

1.3.5 Units of rate constants

Unitary rate constants obey the dimensional relationship $(\text{mol}\,L^{-1})^{-(n-1)}\,s^{-1}$, where n is the *order* of the reaction. As shown in the worked example in Section 1.3.1, a unitary rate constant is denoted by a lowercase k with a subscripted integer that identifies the particular reaction in a system of reactions (e.g., k_{-1}). The direction of a reaction is usually written with an arrow facing from left to right on the page with a positive sign (or no sign) on the subscripted number of the rate constant indicating that is applies to the forward reaction. The negative subscript refers to the reverse reaction that is indicated by the arrow drawn from right to left.

Q: What is the mathematical expression that describes the rate at which A is converted into P, in the following first-order reaction?

$$A \xrightarrow{k} P \quad. \tag{1.8}$$

A: The rate equation is

$$\frac{d[A]}{dt} = -k[A] \quad. \tag{1.9}$$

On the left-hand side of the differential rate-equation, Eqn [1.8], the units (or dimensions) are those of reaction rate, namely, $\text{mol}\,L^{-1}\,s^{-1}$, and those on the right must match these, according to the basic tenet of dimensional analysis. The [A] term has units of $\text{mol}\,L^{-1}$ so k must have units of s^{-1}. Thus, simple dimensional analysis leads directly to the general expression (see above) for the units of a rate constant.

1.3.6 Extent of reaction

This is a useful notion that expresses the fractional departure of a reaction from its commencement to the situation at a particular time. It is a dimensionless ratio that can apply to the decline of a substrate or the appearance of a product. In the latter case, if [P] increases, the extent of the reaction is given by $[P]_t/[P]_\infty$, where t denotes the time at which the extent of the reaction is determined, and ∞ implies the concentration of P that pertains at a very long time when the system is no longer changing. For a substrate A which decreases with time, the expression for the extent of the reaction at time t is

$([A]_0 - [A]_t)/([A]_0 - [A]_\infty)$, where the subscript 0 specifies the concentration of A at zero-time.

1.4 Time Courses of Reactions

1.4.1 Introduction

The chemical reaction depicted in the Section 1.3.5 has an associated differential equation that describes the rate of the reaction for any given concentration of the reactant A. The time course of [A] is described by a mathematical equation that is obtained by solving this differential equation; in other words, we integrate the differential equation to obtain [A] as a function of time. The solution of Eqn [1.8] is a fundamental result in integral calculus and is

$$[A] = [A]_0 \, e^{-k\,t} \ . \tag{1.10}$$

Q: What is the shape of the graph of Eqn [1.10]?

A: To plot the equation we must choose the *initial condition*, namely, the value of the dependent variable [A] at t = 0, and also assign a value to *k*. However, we must first define the function. This is done with the following *Mathematica* input:

```
a[t_] := a[0] e^-k t;
```

There are four important features to note about this definition. The first is the use of lower case 'a' rather than 'A' to denote the concentration of the reactant. Because *Mathematica* commands all begin with a capital letter, it is good practice to use lower case letters when defining our own variable- and function-names. This allows *Mathematica* commands and functions to be easily distinguished from user-defined ones in programs. The second point is that a[0] denotes the concentration of A at t = 0. Third, the definition involves the use of the _ (called a "blank") on the left-hand side *(lhs)* of the equation. This indicates that the argument of the function can be any symbol, but the symbol must also appear on the right-hand side *(rhs)* in the definitional equation. This ability to define so easily our own functions is an important feature of *Mathematica*. The fourth point to note is the use of the symbol := to assign the *lhs* of the function to the *rhs*. The colon before the equals sign indicates that this assignment is a delayed assignment, which means that the *rhs* of the equation is not evaluated when the assignment is made, but instead it is evaluated each time the value of *lhs* is requested in the program. In contrast an equals sign without the colon invokes an immediate assignment. Thus *lhs = rhs* evaluates *rhs* at the time the assignment is made in the program. In the program below we use = to assign the initial condition to A and to assign the value to *k*.

```
a[0] = 10;
k = 0.5;
```

Finally, we use the function **Plot** to draw the graph of the expression from t = 0 to t = 10.

```
gph1 = Plot[a[t], {t, 0, 10},
   AxesLabel -> {"Time ", "Concentration"} ];
```

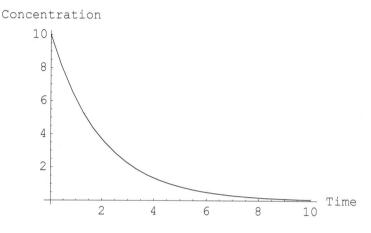

Figure 1.1. Time course of the so-called exponential decay defined by Eqn [1.10].

To determine the value of **a[t]** at any time, say, t = 4, we simply type the following:

```
a[4]
```

```
1.35335
```

$f[x\_] := x^2$	define the function $f(x) = x^2$
$lhs = rhs$	*rhs* is evaluated when the assignment is made
$lhs := rhs$	*rhs* is evaluated each time the value of *lhs* is requested
Plot[f, {x, *xmin*, *xmax*}, *option* $->$ *value*]	plot f as a function of x from *xmin* to *xmax*. See *Mathematica* help browser for a list of *options*

Basic *Mathematica* commands.

Q: Use the *Mathematica* function DSolve to verify that Eqn [1.10] is indeed the solution to the differential equation defined by Eqn [1.9].

A: **DSolve** is a powerful function that determines analytical solutions to differential equations. Before we use **DSolve** it is useful to unassign or clear those variables and names we have used in the previous worked example. These assignments are permanent in a *Mathematica* session and are kept until they are explicitly removed, or when we quit the *Mathematica* session. By clearing values we eliminate a common source of error that arises when programming in *Mathematica*. In fact, we find it good programming practice to clear all variables that we intend to use in a program in the first line of the code, just in case these variables have been used previously and still have assigned values.

```
Clear[a, k]
```

By solving Eqn [1.9] with the initial condition $[A]_0 = A0$ we obtain the required result.

```
solution = DSolve[{a'[t] == -k a[t], a[0] == a0}, a[t], t]
```

$\{\{a[t] \rightarrow a0\ e^{-kt}\}\}$

There are two important points to note about **DSolve**. First, the equations and initial conditions must be written as *lhs == rhs* where == is the logical operator equals sign which is different from the immediate assignment (=) and delayed assignment (:=) operators. The second is that the solution to the differential equation is given as a replacement rule rather than a simple mathematical expression. Replacement rules along with the **/.** (**ReplaceAll**) command can be used to transform any expression. This subtle and very valuable operator is illustrated next.

```
a[t] /. solution
```

$\{a0\ e^{-kt}\}$

The replacement rule can be converted to a function by using the following input:

```
a[t_] = a[t] /. solution;
```

Hence we can plot **a[t]** by using

```
k = 0.5;
a0 = 10;
Plot[a[t], {t, 0, 10},
  AxesLabel -> {"Time ", "Concentration"}];
```

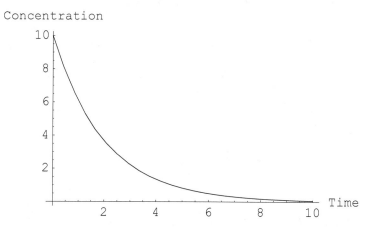

Figure 1.2. Time course of a reaction defined by Eqn [1.10] with the initial condition $[A]_0 = 10$ mmol L^{-1} and rate constant k = 0.5. The ordinate denotes concentration in units of mmol L^{-1} and the abscissa denotes time in units of seconds.

We can also determine the value of [A] at any time, say t = 4, by using the following input:

```
a[4]
{1.35335}
```

$x =$. or **Clear**[x]	remove any value assigned to x
DSolve[$eqns$, $f[x]$, x]	solves a differential equation for the function $f[x]$, with independent variable x
$==$	the logical operator equal
$lhs \rightarrow rhs$ or $lhs \rightarrow rhs$	$lhs \rightarrow rhs$ represents a rule that transforms lhs to rhs
$expr$ /. $lhs \rightarrow rhs$	apply a transformation rule to $expr$
$f[x\_] = f[x]$ /. $rule$	define a function using a repacement rule

More *Mathematica* commands.

1.4.2 Half-life

The time taken for [A] to fall to half of its initial value is called the *half-life* of the reaction. This time is a function of k, as can be seen from the following:

$$\frac{[A]}{[A]_0} = \frac{1}{2} = e^{-k\,t_{1/2}} \ , \qquad\qquad [1.11]$$

hence,

$$\text{Log}[0.5] = -k\,t_{1/2} \ , \qquad\qquad [1.12]$$

so

$$t_{1/2} = \text{Log}[2]/k = 0.693/k \ . \qquad\qquad [1.13]$$

1.4.3 Reaction lifetime

An alternative term to that of half-life is the *lifetime* of the reaction; it is denoted by τ. Its value is the reciprocal of the first order rate constant, i.e., $\tau = 1/k$. We gain further insight into the meaning of τ from Eqn [1.10]:

$$\frac{[A]}{[A]_0} = e^{-k\,1/k} = e^{-1} \ . \qquad\qquad [1.14]$$

In words, τ is the time taken for [A] to fall to $1/e$, or 0.368, of its initial value.

Q: What are the values of the half-life and lifetime of the reaction shown in the previous question? Draw a solid vertical line at the half-life, and a dashed line at the lifetime, on the decay curve.

A: Since $k = 0.5$ s^{-1}, then from Eqn [1.13], $t_{1/2} = 0.693/0.5 = 1.39$ s, and from Section 1.4.3, $\tau = 1/0.5 = 2$ s. Notice that the lifetime is longer than the half-life by a factor of $1/\text{Log}[2]$, or 1.44.

The requisite graph is generated using the following functions.

First we define the lines to be drawn. These are generated using the **Graphics** function of which more details can be found in the *Mathematica* help browser.

```
halfLifeLine = Graphics[
    {AbsoluteThickness[2], Line[{{1.39, 0}, {1.39, 10}}]}];
lifeTimeLine = Graphics[{Dashing[{0.05, 0.02, 0.05, 0.02}],
    Line[{{2.0, 0}, {2.0, 10}}]}];
```

The lines are displayed together with the plot of Eqn [1.10] by using the **Show** function. Recall that we assigned the plot of Eqn [1.10] to the name **gph1** in the previous example.

```
Show[{gph1, halfLifeLine, lifeTimeLine},
    AxesLabel -> {"Time ", "Concentration"}];
```

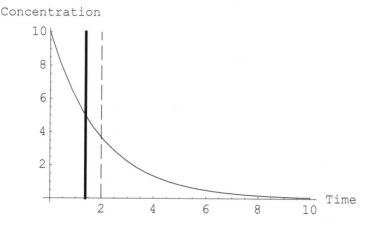

Figure 1.3. Exponential decay with $[A]_0$ = 10 mmol L^{-1} and k = 0.5 s^{-1}, showing the value of the half-life of the reaction (vertical solid line) and the lifetime of the reaction (vertical dashed line), respectively.

Graphics[*primatives*, *option* −>*value*]	represents a two-dimensional graphical image. See help browser for a list of primatives and options
Show[*g₁*, *g₂*, ... , *option* −>*value*]	shows several plots and graphics objects combined

Dealing with graphics.

1.4.4 Coupled first-order reactions

As a prelude to modelling sequences of enzyme-catalyzed reactions, consider a sequence of only two reactions. This configuration is said to be a *coupled* reaction.

$$A \overset{k_1}{\to} B \overset{k_2}{\to} C \; . \tag{1.15}$$

From Section 1.3.1, and by analogy with Eqn [1.10], the rate equations that describe the kinetics of this coupled system are

$$\frac{d[A]}{dt} = -k_1[A] \quad , \tag{1.16}$$

$$\frac{d[B]}{dt} = +k_1[A] - k_2[B] \quad , \tag{1.17}$$

$$\frac{d[C]}{dt} = +k_2[B] \quad .$$

[1.18]

Q: Is it possible to solve the system of differential equations defined by Eqns [1.16 - 1.18] by using the function DSolve?

A: Yes, the program is as follows:

Use **DSolve** to solve the first equation .

```
Clear[a, b, c, a0, b0, c0, Subscript];

firstDE = DSolve[{a'[t] == -k₁ a[t], a[0] == a0}, a[t], t]

{{a[t] → a0 e^(-t 0.5₁)}}
```

Use this solution to define a function for the time course of **a[t]**.

```
a[t_] = a[t] /. firstDE;
```

Next, solve the second and third differential equations and define the appropriate functions for **b[t]** and **c[t]**.

```
secondDE = DSolve[{b'[t] == k₁ a[t] - k₂ b[t], b[0] == b0}, b[t], t]
b[t_] = b[t] /. secondDE;

thirdDE = DSolve[{c'[t] == k₂ b[t], c[0] == c0}, c[t], t]
c[t_] = c[t] /. thirdDE;
```

$$\left\{\left\{b[t] \rightarrow \frac{1}{0.5_1 - 0.5_2} \right.\right.$$
$$\left(e^{-t\,(0.5_1-0.5_2)-t\,0.5_2}\,(-a0\,0.5_1 + a0\,e^{t\,(0.5_1-0.5_2)}\,0.5_1 +\right.$$
$$\left.\left.\left. b0\,e^{t\,(0.5_1-0.5_2)}\,0.5_1 - b0\,e^{t\,(0.5_1-0.5_2)}\,0.5_2)\right)\right\}\right\}$$

$$\left\{\left\{c[t] \rightarrow \right.\right.$$
$$\frac{1}{0.5_1 - 0.5_2}\,(e^{-t\,0.5_1}\,(a0\,e^{t\,0.5_1}\,0.5_1 + b0\,e^{t\,0.5_1}\,0.5_1 + c0\,e^{t\,0.5_1}\,0.5_1 -$$
$$a0\,e^{t\,(0.5_1-0.5_2)}\,0.5_1 - b0\,e^{t\,(0.5_1-0.5_2)}\,0.5_1 + a0\,0.5_2 - a0\,e^{t\,0.5_1}$$
$$\left.\left.\left. 0.5_2 - b0\,e^{t\,0.5_1}\,0.5_2 - c0\,e^{t\,0.5_1}\,0.5_2 + b0\,e^{t\,(0.5_1-0.5_2)}\,0.5_2)\right)\right\}\right\}$$

Hence, after having defined the parameter values and initial conditions, we plot the time courses of **a[t]**, **b[t]**, and **c[t]** as follows:

```
a0 = 10; b0 = 0; c0 = 0;

k₁ = 0.5; k₂ = 0.6;

Plot[{a[t], b[t], c[t]}, {t, 0, 10},
   AxesLabel -> {"Time ", "Concentration"}];
```

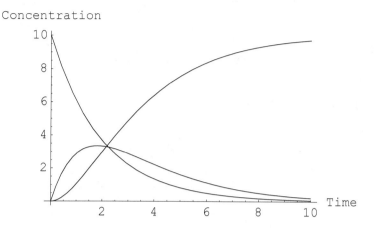

Figure 1.4. Simulation of the time course of the reaction scheme described by Eqn [1.15] using DSolve, with the differential equations in Eqns [1.16 - 1.18]. The ordinate denotes concentration in mmol L^{-1} and the abscissa, time in seconds.

In the previous worked example each differential equation was solved sequentially, with the solution of the first giving the expression for the dependent variable **[A]**. This was then fed into the analysis for **[B]** and so on. Performing the integration in sequence like this was done only for pedagogical purposes. However, it is possible with the function **DSolve** to enter a complete list that contains all the relevant expressions for the differential equations and initial conditions, and to solve them simultaneously. This is done as follows, where we use a different set of parameter values from the example above so that different time courses are obtained.

Q: Can the differential equations in the previous question be solved simultaneously using DSolve?

A: Yes, **DSolve** can be used in the following way:

```
Clear[a, b, c, a0, b0, c0, Subscript];

solution = DSolve[
    {a'[t] == -k₁ a[t],
     b'[t] == k₁ a[t] - k₂ b[t],
     c'[t] == k₂ b[t],
     a[0] == a0,
```

```
      b[0] == b0,
      c[0] == c0},
     {a[t], b[t], c[t]}, t]
```

$$\Big\{\Big\{a[t] \to a0\ e^{-t\ 0.5_1},\ b[t] \to \frac{1}{-0.5_1 + 0.5_2}\ (e^{-t\ 0.5_1 - t\ 0.5_2}$$

$$(-a0\ e^{t\ 0.5_1}\ 0.5_1 - b0\ e^{t\ 0.5_1}\ 0.5_1 + a0\ e^{t\ 0.5_2}\ 0.5_1 + b0\ e^{t\ 0.5_1}\ 0.5_2)),$$

$$c[t] \to \frac{1}{-0.5_1 + 0.5_2}\ (e^{-t\ 0.5_1 - t\ 0.5_2}\ (a0\ e^{t\ 0.5_1}\ 0.5_1 + b0\ e^{t\ 0.5_1}\ 0.5_1 -$$

$$a0\ e^{t\ 0.5_1 + t\ 0.5_2}\ 0.5_1 - b0\ e^{t\ 0.5_1 + t\ 0.5_2}\ 0.5_1 - c0\ e^{t\ 0.5_1 + t\ 0.5_2}\ 0.5_1 -$$

$$b0\ e^{t\ 0.5_1}\ 0.5_2 - a0\ e^{t\ 0.5_2}\ 0.5_2 + a0\ e^{t\ 0.5_1 + t\ 0.5_2}\ 0.5_2 +$$

$$b0\ e^{t\ 0.5_1 + t\ 0.5_2}\ 0.5_2 + c0\ e^{t\ 0.5_1 + t\ 0.5_2}\ 0.5_2))\Big\}\Big\}$$

By specifying the appropriate functions, parameter values, and initial conditions, the results of the integration can be plotted.

```
     a[t_] = a[t] /. solution;
     b[t_] = b[t] /. solution;
     c[t_] = c[t] /. solution;

     a0 = 10; b0 = 0; c0 = 0;

     k₁ = 0.5; k₂ = 0.3;

     Plot[{a[t], b[t], c[t]}, {t, 0, 10},
        AxesLabel -> {"Time ", "Concentration"}];
```

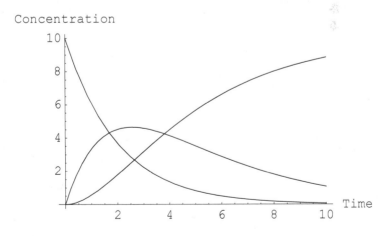

Figure 1.5. Simulation of the time course of the reaction scheme described by Eqn [1.15] using DSolve with the differential equations in Eqns [1.16 - 1.18]; but unlike the analysis for Fig. 1.3, the equations were solved simultaneously. The ordinate denotes concentration in mmol L^{-1} and the abscissa, time in seconds.

1.4.5 Multiple-coupled reactions

What if the reaction sequence is more extensive? Is it possible to solve the simultaneous differential equations analytically? The answer is "Yes!" but only in general if the reactions are simple first-order ones. Indeed *Mathematica* solves these with ease. Consider, for example, the sequence of five reactions with six reactants:

$$A \overset{k_1}{\to} B \overset{k_2}{\to} C \overset{k_3}{\to} D \overset{k_4}{\to} E \overset{k_5}{\to} F \quad . \tag{1.19}$$

Integrating the six differential equations 'by hand' is very tedious (e.g., reference[2]), but *Mathematica* readily yields the exact analytical solutions.

Q: Derive and solve the system of differential equations resulting from the reaction scheme shown in Eqn [1.19].

A: Use **DSolve** in the same manner as for the previous example.

```
Clear[a, b, c, d, e, f, a0, Subscript];

solution = DSolve[{
    a'[t] == -k₁ a[t],
    b'[t] == k₁ a[t] - k₂ b[t],
    c'[t] == k₂ b[t] - k₃ c[t],
    d'[t] == k₃ c[t] - k₄ d[t],
    e'[t] == k₄ d[t] - k₅ e[t],
    f'[t] == k₅ e[t],
    a[0] == a0, b[0] == 0, c[0] == 0, d[0] == 0, e[0] == 0, f[0] == 0},
    {a[t], b[t], c[t], d[t], e[t], f[t]}, t];
```

And, with the following initial conditions, we can plot the results.

```
a0 = 10;

k₁ = 0.5; k₂ = 0.55; k₃ = 0.6; k₄ = 0.65; k₅ = 0.7;

Plot[Evaluate[{a[t], b[t], c[t], d[t], e[t], f[t]} /. solution],
    {t, 0, 10}, PlotRange → All,
    AxesLabel -> {"Time ", "Concentration"}];
```

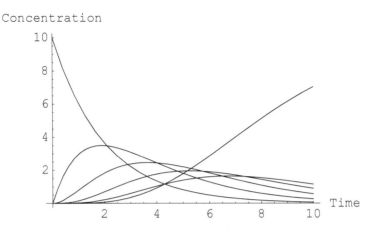

Figure 1.6. Simulation of the time course of the reaction scheme described by Eqn [1.19] using DSolve with the corresponding differential rate equations. The ordinate denotes concentration in mmol L^{-1} and the abscissa, time in seconds.

Notice that in the solution to this problem we use a different method for evaluating and plotting the results generated by **DSolve**. Instead of using the replacement rule generated by **DSolve** to define new functions, we use a combination of the **Evaluate** function and the **/.** operator directly within the **Plot** function itself. This is often a more efficient way of using results which are expressed as replacement rules. In order to evaluate the results of a replacement rule at a specific time (say, t = 4) without defining a new function, we can use the following commands:

```
{a[t], b[t], c[t], d[t], e[t], f[t]} /. solution /. t -> 4

{{1.35335, 2.45321, 2.44581, 1.7734, 1.04475, 0.929452}}
```

Plot[Evaluate[$f[x]$ /. *rule*], {*x, xmin, xmax*}]	Plotting solutions given as replacement rules
$f[x] = f[x]$ /. *rule* /. *x --> value*	Evaluating solutions given as replacement rules at particular value of the independent variable

Using solutions expressed as replacement rules.

From the above example we see that the analytical expression for each of the reactants in the scheme in Eqn [1.19] is a function of time which is mathematically described as a *sum of exponentials*. These complicated expressions can be viewed by removing the end-of-line semicolons and re-executing the first Cell in the example. In a linear reaction system like that above, the exact number of exponentials in the expression for the concentration of a species is equal to 1 + the number of species that precedes it in the sequence. The *exponents* are a complicated series of sums of various products of the rate constants, and they are preceded by similarly complicated *pre-exponential* coefficients. The complexity evident in such analytical solutions does not bode well for a direct analytical assault on the kinetic representation of a metabolic system with hundreds of reactants!

1.4.6 Non-first-order systems

When a reaction scheme contains second- and higher-order reactions (see Section 1.3.4) it turns out that except in very special cases, there is no general analytical solution to the set of simultaneous differential equations that describe the system. When a differential equation contains the product of two or more dependent variables (concentrations of reactants in the present context) it is said to be *nonlinear*. For example, the next, relatively simple, reaction scheme cannot be solved analytically because two of its differential equations are nonlinear:

$$ A + B \underset{k_{-1}}{\overset{k_1}{\rightleftharpoons}} C \overset{k_2}{\rightarrow} D \ . \tag{1.20} $$

However, the equations can be solved by using a procedure called *numerical integration*. This is the topic of the next section. In order to provide some motivation to proceed with the present topic it is encouraging to note that in *Mathematica,* numerical integration of an array of linear or nonlinear differential equations is done by simply adding the letter "N" to the **DSolve** function, viz., **NDSolve**.

NDSolve[*eqns* ,   find numerical solutions for several functions
{ y_1, y_2, ...}, {*x*, *xmin*, *xmax*}] y_i with *x* in the range *xmin* to *xmax*

Finding numerical solutions to differential equations.

Q: Simulate the time course of the reaction scheme in Eqn [1.20] using NDSolve.

A: When performing numerical integration, the output is given as numerical values, so all initial conditions and parameter values must be specified before executing the function or algorithm. There can be no parameter with an unassigned value and this includes specifying the minimum and maximum values of the independent variable, time.

```
Clear[a, b, c, d];

a0 = 5;
b0 = 4.5;
k₁ = 1;
k₋₁ = 1;
```

```
k₂ = 1;

solution = NDSolve[{
    a'[t] == -k₁ a[t] b[t] + k₋₁ c[t],
    b'[t] == -k₁ a[t] b[t] + k₋₁ c[t],
    c'[t] ==   k₁ a[t] b[t] - k₋₁ c[t] - k₂ c[t],
    d'[t] ==   k₂ c[t],
    a[0] == a0,
    b[0] == b0,
    c[0] == 0,
    d[0] == 0},
    {a[t], b[t], c[t], d[t]},
    {t, 0, 10}]
{{a[t] → InterpolatingFunction[{{0., 10.}}, <>][t],
  b[t] → InterpolatingFunction[{{0., 10.}}, <>][t],
  c[t] → InterpolatingFunction[{{0., 10.}}, <>][t],
  d[t] → InterpolatingFunction[{{0., 10.}}, <>][t]}}
```

Notice that unlike **DSolve**, where the solutions for each reactant were given as analytical functions, the solutions for the reactants resulting from **NDSolve** are given as objects called **Interpolating Functions**. The **Interpolating Function** effectively stores a table of values for each reactant as a function of t, then interpolates in this table to find an approximation to each reactant at the particular t that is requested.

We can plot the results in exactly the same way as we did for **DSolve**.

```
Plot[Evaluate[{a[t], b[t], c[t], d[t]} /. solution],
    {t, 0, 10}, AxesLabel -> {"Time ", "Concentration"}];
```

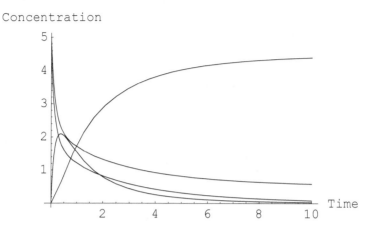

Figure 1.7. Simulation of the time course of the reaction described by Eqn [1.20], using **NDSolve** operating on the corresponding differential rate equations. The ordinate denotes concentration in mmol L^{-1} and the abscissa, time in seconds.

1.5 Numerical Integration of Differential Equations

1.5.1 General

It was our intention in presenting Section 1.4 to provide the motivation for the use of numerical analysis, as opposed to analytical integration, for modelling the dynamics of complicated reaction schemes. It will become obvious that numerical integration can be a robust and reliable means of solving arrays of nonlinear differential equations. The only disadvantage of this approach is that an analytical or general expression is not obtained, only a series of numbers expressed as **InterpolatingFunction** objects in *Mathematica*. Hence a feeling for how changes in parameter values affect a solution really only emerges from repeated simulations rather than from inspecting and manipulating an analytical expression. On the other hand, numerical integration turns out to be the only tractable computational method for modelling time courses of schemes with myriad reactants. Therefore, we proceed by developing an understanding of this part of numerical analysis to a level that is sufficient for our current purposes.[3]

We concentrate first on solving a single first-order ordinary (with no partial derivatives) differential equation that has one initial condition. Importantly, the methods that we develop can be easily extended to arrays of simultaneous differential equations. In general, the single differential equation that describes a simple first-order reaction is expressed as

$$y' = f[t, y] \quad , \tag{1.21}$$

where the prime denotes the first derivative, and the initial condition is

$$y[t_0] = y_0 \quad . \tag{1.22}$$

Equation [1.21] can be viewed as the specification of a family of curves for which the *slope* at any point (t, y) is given by the formula f[t, y]; and Eqn [1.22] ties down the actual function, y[t], to being one particular member of the family of curves. In other words, a *solution* of Eqns [1.21] and [1.22] is defined as the expression for y given in terms of various parameters, as well as t, that satisfies both of the equations. This is illustrated with the three functions in the next example. If we declare that the initial condition is y[0] = 1, then the particular function that we have identified is the one labelled y_2 in the example below.

Q: What is the effect on the shape, or the relative position, of the exponential curves that are solutions of Eqn [1.21], brought about by changing the value of the pre-exponential coefficient?

A: The plots are as follows:

```
y₁[t_] := 2 eᵗ;
y₂[t_] := eᵗ;
y₃[t_] := 0.5 eᵗ;

Plot[{ y₁[t], y₂[t], y₃[t]},
  {t, -3, 3}, PlotRange → {{-3, 3}, {0, 6}},
  AxesLabel → {"t", "              y       y₁  y₂   y₃"}];
```

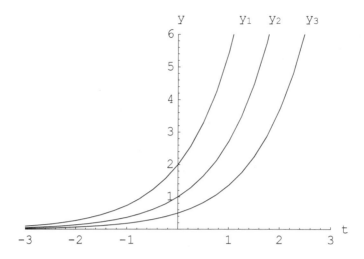

Figure 1.8. Plots of the function y = A e^t for three different values of A: 2, 1, and 0.5, respectively.

1.5.2 Numerical integration – overview

A preliminary view of how we obtain a numerical solution for y[t] is as follows. We first substitute a value of t into f[t, y] in Eqn [1.21] and thus determine the slope of the curve. At the outset, the only value of y[t] that is known is the one at the point (t_0, y_0); so we begin our solution here. We compute the slope of the curve at t = t_0 and then draw a short tangent line through (t_0, y_0). If we denote the increment in t by h, we arrive at the new coordinate $t_1 = t_0 + h$. This value of t, coupled with the equation for a straight line [y = $y_0 + (t_1 - t_0)$ slope], yields the new value y = y_1. We then substitute the new values y_1 and t_1 into the slope formula to obtain f[t_1, y_1]. By cycling through (called iterating) this process, we obtain a series of joined straight-line segments (i.e., a portion of a polygon) that approximates the true function. This is called Euler's point-slope method of numerical integration of a differential equation.

Consider the implementation of the Euler method to the exponential function.

Q: Use Euler's method to solve y'[t] = y[t] with y[0] = 1 and compare this solution graphically with the analytical solution.

A: First, we use **DSolve** to generate an analytical solution to this differential equation in the usual manner.

```
solution =
 DSolve[{y_analytical '[t] == y_analytical[t], y_analytical[0] == 1},
  y_analytical[t], t]
y_analytical[t_] = y_analytical[t] /. solution;

{{y_analytical[t] → e^t}}
```

To solve this differential equation by Euler's method we begin by specifying the initial condition and choosing a stepsize, say, h = 0.5.

```
y_em[0] = 1;
h = 0.5;
```

Now apply Euler's method iteratively using the **Do** command for 5 iterations, and then put the results into a **Table** of ordered pairs (t, y[t]).

```
Do[y_em[i] = y_em[i - 1] + h y_em[i - 1], {i, 1, 5}];
emTable = Table[{i h, y_em[i]}, {i, 0, 5}]

{{0, 1}, {0.5, 1.5}, {1., 2.25},
 {1.5, 3.375}, {2., 5.0625}, {2.5, 7.59375}}
```

Graphically compare the two results using **ListPlot** to plot the table of points in **emTable** and the familiar **Plot** function to graph the analytical solution. Notice that in the following functions we suppress the initial output of the **ListPlot** and **Plot** commands by using the option **DisplayFunction→Identity**; this allows us to produce a single combined plot using the **Show** function.

```
gph1 = ListPlot[emTable, PlotStyle -> {PointSize[0.025]},
    DisplayFunction → Identity];
gph2 = ListPlot[emTable, PlotJoined → True,
    DisplayFunction → Identity];
gph3 = Plot[y_analytical[t], {t, 0, 2.5},
    DisplayFunction → Identity];

Show[gph1, gph2, gph3,
    DisplayFunction → $DisplayFunction, AxesLabel → {"t", "y"}];
```

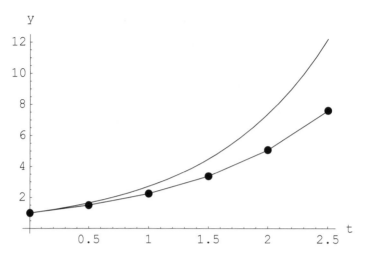

Figure 1.9. The analytical solution of Eqn [1.20] and the numerical solution (large points) obtained by using Euler's point-slope method.

Do[*expr*, {*i, imin, imax, di*}]	evaluate *expr* with *i* running from *imin* to *imax* in steps of *di*. If *di* is omitted, step size is 1
Table[*expr*, {*i, imin, imax, di*}]	make a list of the values of expr with *i* running from *imin* to *imax* in steps of *di*. If *di* is omitted, step size is 1
ListPlot[*list*, *option* —> *value*]	plot points (x_1, y_1), ... in *list*. See help browser for a list of *options*

Loops, lists, and plotting lists.

It can be seen from Fig. 1.9 that there are dangers with approximating a smoothly curved solution by a series of linear segments. So, it is important to find ways of taking into account the likely curvature of the real solution, which of course we don't know *a priori*. It is a marvel of modern numerical analysis that error-knowing-and-minimizing algorithms have been developed. To this end, there are two basic categories of numerical integration: (1) the *single step* methods that rely on information about the curve, one point at a time. They are said to be the *direct* methods as they do not iterate a solution (see much more on this topic below); and (2) *multi-step* methods in which each successive point on the curve is found with fewer evaluations of the slope function (derivative) than in (1); but iterations are required to arrive at a sufficiently accurate value. Most methods of this type are called *predictor-corrector* methods. One of these is the default algorithm used in **NDSolve** in *Mathematica*. There is a problem with getting these methods started, but this disadvantage is offset by the side benefit of obtaining an estimate of the error as the solution proceeds. In practice with *Mathematica* the peculiarities of the various methods are transparent to the user, and the most modern algorithm is implemented unless the default is overridden by specific commands (see the *Mathematica* book for details).

1.5.3 Taylor series solution

A systematic study of numerical integration is usefully begun with the Taylor series expansion of a mathematical function; this theory is often discussed in intermediate-level mathematics. The resulting power-series provides a representation of an arbitrary function at a given point t when its value is known in the neighborhood of the point t_0. The function must be continuous or at least piecewise continuous and be infinitely differentiable in order for this representation to be strictly meaningful. Hence,

$$y[t] = y[t_0] + y'[t_0] h + y''[t_0] h / 2! + y'''[t_0] h / 3! + \dots y^{(n)}[t_0] h^n / n! + \dots \qquad [1.23]$$

where we define a step size $h = (t - t_0)$, and $y^{(j)}[t_0]$ denotes the j-th derivative of y[t] evaluated at $t = t_0$. The Taylor series approximation is used to generate the value of y at $t = t_0 + h$. Then the value at $t = t_0 + 2h$ is generated by replacing t_0 in Eqn [1.23] by $t = t_0 + h$. We repeat the application of the Taylor series with increasing values of t, using the fixed step size h. This means that the (m+1)-th value of y is y[t_{m+1}] and it is based on the value at t_m via Eqn [1.23]. Hence,

$$y[t_{m+1}] = y[t_m] + y'[t_m] h + y''[t_m] h / 2 + y'''[t_0] h / 3! + \dots y^{(n)}[t_m] h^n / n! + \dots \qquad [1.24]$$

where, in general, $h = (t_{m+1} - t_m)$. Henceforth, we adopt the simpler convention that $y_m \equiv y[t_m]$.

We can approximate any function by truncating the infinite Taylor series after m terms; and the more terms we include, the more accurate is the approximation. For the first term of the series we already have an expression for the *slope* of the function (Eqn [1.21]) but for each successive term a differentiation operation is required. Hence,

$$y'_m = f[t_m, y_m] \ . \qquad [1.25]$$

And, y" is obtained by taking the second derivative with respect to both t and y:

$$y'' = \frac{\partial f}{\partial t} + f \frac{\partial f}{\partial y} \ , \qquad [1.26]$$

or, more concisely,

$$y'' = f_t + f \ f_y \ , \qquad [1.27]$$

where differentiation with respect to t or y is indicated by the subscript on f. Note that when carrying out this analysis, the differentiation is done first, and then the actual numerical value is obtained for each term by substituting the values of t_m and y[t_m] into them.

By using this nomenclature, Eqn [1.23] becomes

$$y_{m+1} = y_m + h\left(f + \frac{h}{2}\,(f_t + f\,f_y)\right) + O(h^3) \ , \tag{1.28}$$

where $O(h^3)$ is read as 'of the order of h cubed,' and it denotes the fact that all subsequent terms contain h to the third or higher power, or degree. In other words, if we use Eqn [1.28] without the $O(h^3)$ term, the error introduced by truncating the series is approximately $K\,h^3$, where K is some constant.

The Taylor series solution of a differential equation is classified as a *one-step* method because the value of y_{m+1} is obtained by using only t_m and y_m, in one step.

A major practical problem with the Taylor series solution is that it is often difficult to obtain the analytical expressions for the second- and higher-order partial derivatives. However, this is easy today with *Mathematica;* and the Taylor series method can provide a means of comparison of other methods of approximation even when they do not require the calculation of higher-order derivatives, as is shown below.

1.5.4 Runge-Kutta methods – overview

These methods constitute a broad class of techniques for numerically solving differential equations. The methods have three distinguishing features: (1) they are called *one-step methods* as they require information at the preceding point only; (2) they agree with the Taylor series (Eqn [1.24]) through to order p, which is therefore called the *order of the method*; and (3) they do not require the evaluation of any higher-order derivatives as they use only the given first-derivative expression, f. This latter property makes the methods more useful than the Taylor series method. However, the price paid for not evaluating the higher-order derivatives is to evaluate far more than one value beyond (t_m, y_m), when stepping from (t_m, y_m) to (t_{m+1}, y_{m+1}).

A geometrical representation is useful for understanding these methods, but remember that algebraic analysis is still needed for their ultimate verification. Suppose that the solution y_m at the point $t = t_m$ is known. Then a line is drawn through the point (t_m, y_m) such that its slope is $y_m' = f[t_m, y_m]$. See Fig. 1.9 where the heavy line denotes the exact but unknown solution and the line just described is denoted by L_1. Then we let y_{m+1} be the point where L_1 intersects the ordinate erected at $t = t_{m+1} = t_m + h$. Hence, the equation for L_1 is

$$y = y_m + y_m'\,(t - t_m) \ , \tag{1.29}$$

but

$$y_m' = f[t_m, y_m] \ , \tag{1.30}$$

and

$$t_{m+1} - t_m = h \ , \tag{1.31}$$

so

$$y_{m+1} = y_m + h\,f[t_m,\ y_m] \ . \tag{1.32}$$

The error at $t = t_{m+1}$ is shown in Fig. 1.9 as ϵ, and Eqn [1.32] agrees with the Taylor series through to terms in h, so the truncation error is represented by

$$\epsilon_T = K h^2 \ . \tag{1.33}$$

Note that although the point $(t_m,\ y_m)$ in Fig. 1.10 is drawn lying on the exact (unknown) curve, in practice it is an approximation, so it will not necessarily do so. Equation (1.31) is called Euler's point-slope method and is the oldest and best known numerical method for integrating a differential equation. However, besides having a relatively large truncation error, it can also be quite unstable. In other words, a small error due to numerical round-off, truncation, or inherent in the underlying function f becomes magnified as t increases. Therefore, more accurate methods have been developed.

In passing, note that according to the definition above, Euler's point-slope method is a first-order Runge-Kutta one, since it agrees with the Taylor series representation of the function through to terms in h.

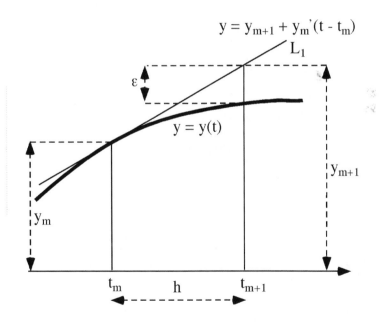

Figure 1.10. Graphical analysis of Euler's point-slope method, which is a first-order Runge-Kutta method of numerical integration.

Euler's point-slope method can be improved in a number of different ways. The two most important ones are called the *Improved Euler method* and the *Modified Euler method* and they are discussed next.

1.5.5 Improved Euler method

With the Improved Euler method (also called Huen's method), we work with the average of the slopes at (t_m, y_m) and $(t_m + h, y_m + h y_m')$. In Fig. 1.11 this point is denoted by (t_{m+1}, y_{m+1}). Geometrically, we use Euler's point-slope method to locate the point $(t_m + h, y_m + h y_m')$ on the line L_1 and at this point we compute the slope of the function leading to line L_2. Then we take the average of the two slopes and obtain the dashed line $\bar{L}$. Finally, the line L is drawn parallel to $\bar{L}$ through the point (t_m, y_m). The point at which this line intersects the ordinate erected at t_{m+1} is taken to be y_{m+1}.

Let us analyze the construction algebraically. The slope of $\bar{L}$ and hence of L is given by

$$\phi[t_m, y_m, h] = \frac{1}{2} \{f[t_m, y_m] + f[t_m + h, y_m + h y_m']\} \quad , \tag{1.34}$$

recalling that

$$y_m' = f[t_m, y_m] \quad , \tag{1.35}$$

so the equation for L is

$$y = y_m + (t - t_m) \phi[t_m, y_m, h] \quad . \tag{1.36}$$

These three equations define the Improved Euler method.

How well does this solution agree with the Taylor series solution? To see this we expand the slope of the function, f, as a Taylor series:

$$f[t, y] = f[t_m, y_m] + (t - t_m) \frac{\partial f}{\partial t} + (y - y_m) \frac{\partial f}{\partial t} + \dots \quad , \tag{1.37}$$

where the partial derivatives are evaluated at $t = t_m$ and $y = y_m$. Substituting $t = t_m + h$ and $y = y_m + h y_m'$ into Eqn [1.36] and using Eqn [1.34] we obtain

$$f[t_m + h, y_m + h y_m'] = f + h f_t + h f f_y + O[h^2] \quad , \tag{1.38}$$

where f and the second-order (partial) derivatives are evaluated at (t_m, y_m). Substituting this expression into the slope formula (Eqn [1.35]) and after some rearrangement we obtain

$$\phi[t_m, y_m, h] = f + \frac{h}{2} \{f_t + f f_y\} + O[h^2] \quad . \tag{1.39}$$

Finally, this expression is used in Eqn [1.36] to provide a direct comparison with the Taylor series:

$$y_m = y_m + h f + \frac{h^2}{2} \{f_t + f f_y\} + O[h^2] \quad . \tag{1.40}$$

Thus from Eqn [1.40] we see that the equation agrees with the Taylor series through terms in h^2. Therefore we say that the Improved Euler method is a second-order Runge-Kutta method. We see that we are required to evaluate the slope $f[t, y]$ twice, at (t_m, y_m) and $(t_m + h, y_m + h y_m')$. In a comparison, the computational effort involved with using the Taylor series is three function (slope) evaluations, f, f_t, and f_y.

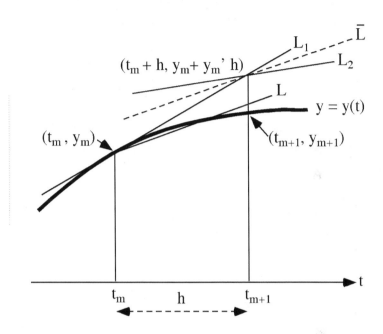

Figure 1.11. Graphical analysis of the Modified Euler method of numerical integration.

1.5.6 Modified Euler method

In the Improved Euler method slopes are averaged, but in this method positions of points are averaged as is shown geometrically in Fig. 1.12. We begin with the line L_1 that passes through (t_m, y_m) and has the slope $f[t_m, y_m]$. We proceed along L_1 and find the point of intersection with the ordinate erected at $t_m + h/2$; this is the point P at which $y = y_m + (h/2) y_m'$. The slope of the function at this point is then calculated:

$$\phi[t_m, y_m, h] = f[t_m + h/2, y_m + (h/2) y_m'] , \qquad [1.41]$$

again, where

$$y_m' = f[t_m, y_m] . \qquad [1.42]$$

The line through P with this new slope is shown as L* in Fig. 1.12. Next we draw a line parallel to L* passing through (t_m, y_m) that is shown as L_0. Now designate the value of y_{m+1} to be the intersection of L_0 with $t = t_m + h$. Hence the equation for L_0 is

$$y = y_m + (t - t_m)\,\phi[t_m, y_m, h] \ . \tag{1.43}$$

Thus,

$$y_{m+1} = y_m + h\,\phi[t_m, y_m, h] \ . \tag{1.44}$$

Equations [1.40 – 1.43] define the modified Euler method that is also known as the improved polygon method, for (almost!) obvious reasons. It can be shown to be a second-order Runge-Kutta method.

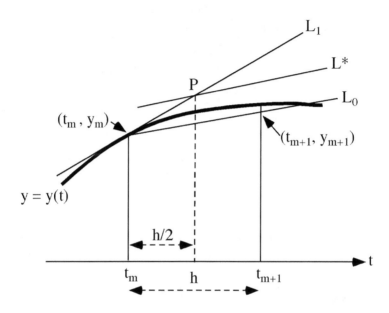

Figure 1.12. Graphical analysis of the Modified Euler method of numerical integration.

1.5.7 General Runge-Kutta methods

Both of the refined Euler methods have a master equation of the form

$$y_{m+1} = y_m + h\,\phi[t_m, y_m, h] \ , \tag{1.45}$$

and in both cases

$$\phi[t_m, y_m, h] = a_1\,f[t_m, y_m] + a_2\,f[t_m + b_1\,h, y_m + b_2\,h\,y_m'] \ . \tag{1.46}$$

So, by using this classification of the methods we can see that the Improved Euler method has the following relationships:

$$a_1 = a_2 = \frac{1}{2} \ , \tag{1.47}$$

$$b_1 = b_2 = 1 \; , \qquad\qquad [1.48]$$

while for the Modified Euler method, we have

$$a_1 = 0, a_2 = 1 \; , \qquad\qquad [1.49]$$

$$b_1 = b_2 = \frac{1}{2} \; . \qquad\qquad [1.50]$$

It can be shown with some clever analysis that a general second-order Runge-Kutta method is one in which, in Eqn [1.46],

$$a_1 = 1 - \omega, a_2 = \omega \text{ and } \omega \neq 0 \; , \qquad\qquad [1.51]$$

$$b_1 = b_2 = \frac{1}{2\,\omega} \; . \qquad\qquad [1.52]$$

The truncation error for any non-zero choice of ω is

$$\varepsilon_T = K\,h^3 \; . \qquad\qquad [1.53]$$

It is mathematically possible to estimate the bounds on $|K|$ and it has been shown that the smallest upper bound is obtained when $\omega = 2/3$.

1.5.8 Higher-order Runge-Kutta methods

Third- and fourth-order Runge-Kutta methods have been developed in ways that are entirely analogous to the second-order ones. Hence we will not work through the derivations but simply state the fourth-order formula:

$$y_{m+1} = y_m + \frac{h}{6} \, (k_1 + 2\,k_2 + 3\,k_3 + k_4) \; , \qquad\qquad [1.54]$$

$$k_1 = f[t_m, y_m] \; , \qquad\qquad [1.55]$$

$$k_2 = f\left[t_m + \frac{h}{2}, y_m + \frac{h}{2}\,k_1\right] \; , \qquad\qquad [1.56]$$

$$k_3 = f\left[t_m + \frac{h}{2}, y_m + \frac{h}{2}\,k_2\right] \; , \qquad\qquad [1.57]$$

$$k_4 = f[t_m + h, y_m + h\,k_3] \; , \qquad\qquad [1.58]$$

and the truncation error is

$$\varepsilon_T = K\,h^5 \; . \qquad\qquad [1.59]$$

1.6 Predictor Corrector Methods

A hallmark of the Runge-Kutta methods is that stepping to the next point, (t_{m+1}, y_{m+1}), uses information at (t_{m+1}, y_{m+1}) but at no *other* prior points. We must evaluate the slope function (derivative) at one or more *subsequent* points, depending on the order of the method. The fact that these methods do not use the accumulated information of prior points, plus the lack of a convenient error estimation procedure, suggest there might be value in devising other methods.

These newer methods turn out to be the Predictor-Corrector ones. As the name implies, a value for y_{m+1} is first *predicted* by one formula and it is then *corrected* by another. If required, the latter value can be *re-corrected* by iteration.

1.6.1 The predictor

A simple approach is as follows. For the predictor we use a simple second-order method, namely,

$$y_{m+1}^{(0)} = y_{m-1} + 2 \, h \, f[t_m, y_m] \; , \qquad\qquad [1.60]$$

where the superscript (0) indicates that this is the initial guess at y_{m+1}, i.e., the predicted value. Immediately note that the method cannot be used to compute y_1 since to do so would require the point y_{-1}. Thus, a Runge-Kutta method is used to predict y_1. Alternatively, we might have thought that Euler's method could have been used here, thus obviating the need for (t_{m-1}, y_{m-1}), but it turns out that the truncation error in this method is routinely too large. The use of prior information on the function leads to the classification of these methods as *multistep* ones.

Figure 1.13 gives some geometrical insight into the operation of the predictor. First, we draw the line L parallel to L_1 through the point $(t_{m-1}, \; y_{m-1})$. This line intersects the ordinate, erected at $t = t_{m+1}$, at the predicted value of $y_{m+1}^{(0)}$. We now improve this prediction by drawing the line L parallel to L_1 through the point $(t_{m-1}, \; y_{m-1})$. This line intersects the ordinate erected at $t = t_{m+1}$ at the predicted value of $y_{m+1}^{(0)}$. Then we improve this prediction.

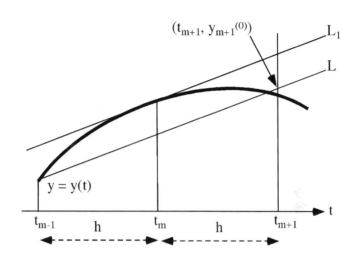

Figure 1.13. Graphical analysis of the predictor part of the predictor-corrector method of numerical integration.

1.6.2 The corrector

Since y_{m+1} is known approximately, the approximate slope at the point $(t_{m+1}, y_{m+1}^{(0)})$ can be calculated. This is shown as the line L_2 in Fig. 1.14. The line L_1 is the same as that in Fig. 1.13 and its slope is given by $f[t_m, y_m]$. Then the slopes of L_1 and L_2 are averaged to give the line L. Finally, the line L_3 is drawn parallel to L through the point (t_m, y_m). Its intersection with the ordinate erected at $t = t_{m+1}$ yields the new approximation to y_{m+1}. This value is called the corrected value $y_{m+1}^{(1)}$. In algebraic terms it is given by

$$y_{m+1}^{(1)} = y_m + \frac{h}{2} \{f[t_m, y_m] + f[t_{m+1}, y_{m+1}^{(0)}]\} \ . \qquad [1.61]$$

Another, and hopefully even better, estimate of $f[t_{m+1}, y_{m+1}]$ is obtained by using $y_{m+1}^{(1)}$ and recorrecting its value. Thus, in general, the i-th approximation to y_{m+1} is given by

$$y_{m+1}^{(i)} = y_m + \frac{h}{2} \left\{f[t_m, y_m] + f\left[t_{m+1}, y_{m+1}^{(i-1)}\right]\right\} \ . \qquad [1.62]$$

Finally, the iteration of this step is stopped when

$$\left| y_{m+1}^{(i+1)} - y_{m+1}^{(i)} \right| < \varepsilon \ , \qquad [1.63]$$

where ε is a value that we specify and which is called the *tolerance*.

It can be shown mathematically that if the partial derivative $\frac{\partial f}{\partial y}$ is bounded and $\leq$

some number M, then if the step-size h is less than 2/M, the solution will converge to a finite value. But it is a curious result that this convergence limit is not necessarily, but almost always, the 'correct' value.

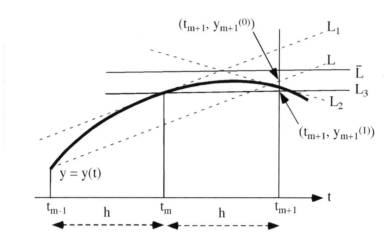

Figure 1.14. Graphical analysis of the corrector part of the predictor-corrector method of numerical integration.

1.6.3 Choosing the value of h

The choice of the initial value of h in the predictor corrector method can be guided by the inequality h < 2/M. However, this is not very convenient because the value of M must be estimated, so almost invariably h is simply chosen to be a small fraction around 0.01% of the maximum time of the simulation. Once the numerical integration has started, the automatic procedures specified above 'kicks in.'

If the value of the expression on the left of Eqn [1.63] does not satisfy the inequality, then h is halved and the predictor step is repeated. This is then followed by the corrector step and its iteration. If the value is smaller than required, h is increased usually by doubling it.

Finally, we must consider the manner in which the truncation error grows; in other words, we must consider the *instability* of the solution. This aspect is addressed by the proper choice of step-size h; and in various *Mathematica* functions the integration algorithm performs this task automatically, so you will not be unduly troubled by this problem.

1.6.4 Numerical integration in *Mathematica* – NDSolve and stiffness

As discussed previously, **NDSolve** is the *Mathematica* function that is used for numerically solving arrays of simultaneous differential equations. The function uses a variety of methods of numerical integration and takes into account an important characteristic known as the 'stiffness' of the array of equations. Stiffness refers to the extent to which the set of equations has members that describe slow processes and members that describe very fast processes. A large range of rate constants means that in order for the numerical integrator to accurately keep track of the fast processes, the step size must be very small; on the other hand, the slow processes could be accurately represented by using very large step sizes. If it is phenomena that occur after a long time that are of interest, the system still has to evolve through the fast stages and this requires time-consuming small step sizes to be used in the integration. A large range of rate constants in a system is referred to as imparting stiffness to the system. The term arose in electrical engineering and it refers to the stiffness of a servo-motor shaft when there is very strong negative feedback in its electrical circuit.

The default method of numerical integration used in **NDSolve** is one based on the LSODE algorithm that detects whether the system of differential equations is stiff or not and uses the most efficient method. For non-stiff systems it uses the method of Adams, while for stiff systems it uses the method of Gear. Because of its use of the Gear algorithm, **NDSolve** is well suited to numerical integration involved in simulating metabolic systems.

Another important feature of **NDSolve** is our ability to specify the required accuracy of its solutions. It will limit the initial and maximum values of the step size h so that the solution will be accurate to a predetermined number of decimal places or a predetermined number of significant figures. Thus, if your goal is to determine the solution to, say, 5 decimal places, you use the option **AccuracyGoal → 5**. While if your goal is 5 significant figures, you use the option **PrecisionGoal → 5**. These options are used in many places throughout the book, e.g., in Section 3.2.4.

1.7 Conclusions

In conclusion, the procedure for writing a set of simultaneous first-order ordinary nonlinear differential equations that describe a biochemical reaction scheme is relatively straightforward. It relies on the principle of mass action. The numerical solution of the equations, however, demands assignment of values to all rate constants and the specification of the initial values of all dependent variables. Under almost all circumstance it is not time efficient to program one's own numerical integration procedure. However, in the spirit of advising against the use of black boxes in science, this chapter has presented the mathematical basis of numerical integration. On the other hand, in the routine practice of computer modelling of enzymic reactions, we advocate the use of packages that have been developed specifically for this purpose; in particular, we aim to convince you of the merits of using *Mathematica* for this task, coupled with the Metabolic Control Analysis (MCA) Package that we introduce in the next chapters. The semantics that need to be adhered to when using *Mathematica* to perform the simulations are described in its Help menu. Only in special circumstances is it useful or indeed possible to derive an analytical solution to the differential equations for a realistic reaction network, so the work described in the next chapters depends on numerical integration.

The derivation of the differential equations that describe very complex metabolic systems, including multistep enzymic reactions, becomes a tedious and error-prone process. Hence the automation of this step is also a welcome development. In the chapters ahead the automatic derivation of the rate equations and may aspects of metabolic simulations are developed in detail.

1.8 Exercises

1.8.1

Modify Eqn [1.15] to make it a pair of reversible reactions,

$$A \underset{k_{-1}}{\overset{k_1}{\rightleftharpoons}} B \underset{k_{-2}}{\overset{k_2}{\rightleftharpoons}} C \ ,$$

and then write the differential equations that describe the kinetics of this reaction scheme.

1.8.2

(1) Derive the time course expressions for each of the reactants in the scheme shown in Exercise 1.1. (2) What do you notice about the number of exponentials in each of the expressions? In other words, for the irreversible system shown in Eqns [1.15] and [1.19], the expression for [A] is a single exponential. (3) Is the rule regarding the number of exponential terms the same in the case of reversible-reactions; if not, why not?

1.8.3

Plot the time course of concentrations for each of A, B, and C, as a function of time using the expressions generated in Exercise 1.1 above. Choose the following parameter values: $A[0] = 10$ mmol L^{-1}, and $k_1 = 1$, $k_{-1} = 1.1$, $k_2 = 1.5$, and $k_{-2} = 1.6$ s^{-1}.

1.8.4

Repeat Exercises 1.8.1 – 1.8.3 above with a larger number of reactants in the linear sequence of reversible reactions and address each of the questions that are posed there.

1.8.5

Rerun the program in the worked example in Section 1.4.6 choosing different values of a0, b0, and the three rate constants. Notice in particular what happens to the maximum value of the concentration of the intermediate species C, and the time at which it occurs, when k_1 and k_{-1} are made large relative to k_2.

It turns out that this reaction scheme, and its solution, are very much like what arise with a simple enzyme. For enzyme systems, the enzyme concentration is usually very low, in the micro- to nanomolar range, but the rate constants that characterize the binding and dissociation steps are large, of the order of $10^6 \, \text{mol}^{-1} \, \text{L}$ and $10^6 \, \text{s}^{-1}$.

1.8.6

Go to the *Mathematica* Help menu and review the attributes of **NDSolve** and **NIntegrate**.

1.9 References

1. Kuchel, P.W. (1985) Kinetic analysis of multienzyme systems in homogeneous systems. In: *Organized Multienzyme Systems* (Welch G.R., Ed.), Academic Press, Orlando. pp. 303-380.
2. Kuchel, P.W. and Ralston, G.B. (1998) *Schaum's Outline of Biochemistry*. 2nd ed., McGraw-Hill, New York, Chapter 9.
3. McCracken, D.D., and Dorn, W.S. (1964) *Numerical Methods and Fortran Programming*. John Wiley & Sons, New York.

2 Elements of Enzyme Kinetics

2.1 Kinetics of Enzymic Reactions

2.1.1 Purpose

In the first chapter the key concepts of chemical kinetics were introduced. All of these concepts are applicable to the chemical reactions which occur inside living organisms. However, a prominent feature of such biochemical reactions is that they tend to be catalyzed by proteins called enzymes. The involvement of enzymes in biochemical reactions adds some special features to the kinetics of biochemical reactions.

The study of biochemical reactions is termed *enzyme kinetics* and in this chapter we present the relevant terms, definitions, and concepts that relate classical chemical kinetics to enzyme kinetics. Thus we will be able to compose the differential equations that are necessary to simulate the time dependence of single and multi-enzyme systems. Since a guiding principle of this book is the *linking of simulations to experimental reality*, the theoretical basis of various methods of data analysis that are used to extract estimates of rate constants in kinetic equations from experimental data are presented; this is done first.

2.1.2 The Michaelis-Menten equation

In 1926 J. B. Sumner[1] reported the crystallization of an enzyme, urease, and thus convinced most chemists and biologists that enzymes are distinct, albeit complex, chemical species that are able to be purified to homogeneity. This concept, coupled with that of Michaelis and Menten in 1913,[2] that enzymes form specific complexes with their reactants, paved the way for a detailed understanding of the chemical mechanisms of individual types of enzymes. One of the simplest experiments that can be carried out on a solution of a particular enzyme is to study the rate at which it converts its substrate(s) to product(s). This process can be studied by using a physical recording device such as a spectrophotometer; the chemical reaction either directly or indirectly develops a chromophore, thus enabling a record of the time dependence of the concentrations of at least one of the reactants. Alternatively, an NMR spectrometer can be used, most often without the requirement for *additional* reactants to generate a detectable chromophore, since almost invariably the NMR spectrum of the product(s) will be different from that of the substrate(s). Experimentally, the effect of substrate concentrations on the rate of the reaction is measured, and because the products of the reaction might inhibit or activate the enzyme as they accumulate during the reaction, or the enzyme may be unstable in the conditions of the assay medium, it is common practice to measure the *initial velocity*, v_0. This initial velocity is measured as the slope of the progress curve of the reaction after extrapolating it back to t = 0. When the substrate concentration, $[A]_0$, is much greater than the enzyme concentration, the overall rate of the reaction is not only proportional to the concentration of the enzyme, but the plot of v_0 versus $[A]_0$ has the form of a rectangular hyperbola. The equation describing the rectangular hyperbola is called the *Michaelis-Menten* equation:

$$v_0 = \left(\frac{d[A]}{dt}\right)_{t=0} = \frac{V_{max}[A]_0}{K_m + [A]_0} \ . \qquad [2.1]$$

This equation has the property that when $[A]_0$ is very large, $v_0 = V_{max}$ (hence it is called the maximum velocity), and when $v_0 = V_{max}/2$, the experimental value of $[A]_0$ is equal to K_m, and this is called the *Michaelis constant*.

Q: Using *Mathematica* construct a plot of the Michaelis-Menten equation over a domain of concentrations of A, for an enzyme having a V_{max} of 1 μmol s^{-1} and a K_m of 1 mM.

A: The following *Mathematica* Cell has the requisite series of functions to perform this task. Recall that items such as the semantics of *delayed evaluation* of equations, and of the **Plot** function, are given in Chapter 1 and also in the *Mathematica* Help Menu.

```
v₀[a_] := Vmax a
          ───────  ;
          Km + a

Vmax = 1 × 10⁻⁶;
Km = 1 × 10⁻³;
Plot[v₀[a], {a, 0, 5×10⁻³},
    PlotRange -> {0, 1×10⁻⁶}, AxesLabel -> {"[A]", "v₀"}];
```

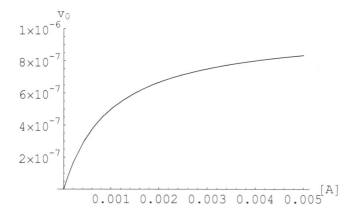

Figure 2.1. The rectangular hyperbolic form of the Michaelis-Menten equation. The ordinate is in units of μmol s^{-1} and the abscissa is in units of mmol L^{-1}.

2.1.3 Graphical evaluation of V_{max} and K_m

The challenge faced by an experimenter is the determination of the two key parameters of the Michaelis-Menten equation, V_{max} and K_m, given a set of data pairs ($[A]_{0,i}$, $v_{0,i}$ i = 1,...,N), where N is usually 5 – 10. Actually, a deeper challenge awaits the experimenter, and that is to determine whether the fitting equation is a realistic description of the data. Nevertheless, suppose that we have established, from inspecting the Michaelis-Menten plot, that the data conform, at least roughly, to a rectangular hyperbola. Then we can make progress with the analysis by rearranging Eqn [2.1] into one of several possible forms that yield straight lines, when the newly transformed data-variables are plotted versus each other. The practical advantages of this mathematical manipulation are that (1) V_{max} and K_m can be determined readily by fitting a straight line to the transformed data; (2) departure of the data from a straight line are more readily detected by eye, than non-conformity to a rectangular hyperbola (these departures may indicate an inappropriateness of the simple model of the enzyme kinetics); and (3) the effects of inhibitors on the reaction can be more easily visualized.

It is worth noting that these data-transformation procedures, while being useful for providing initial estimates of parameters and for 'eye-balling' the data, do bias the error structure of the data and as such yield biased estimates of the kinetic parameters. Thus they have been superseded by non-linear regression for the final or definitive estimates of the parameters and their associated errors (see Chapter 6).

2.1.4 Lineweaver-Burk plot

The most commonly used transformation of Eqn [2.1] entails taking the reciprocal of each side of the equation to yield the Lineweaver-Burk, or double reciprocal, plot:

$$\frac{1}{v_0} = \frac{K_m}{V_{max}} \frac{1}{[A]_0} + \frac{1}{V_{max}} . \qquad [2.2]$$

A plot of the transformed data pairs $(1/[A]_{0,i}, 1/v_{0,i})$ i = 1,..., N, gives a straight line with ordinate and abscissa intercepts at $1/V_{max}$ and $-1/K_m$, respectively.

Q: Generate a Lineweaver-Burk plot for the enzyme described in the previous worked example.

A: First we need to generate a **Table** of ordered pairs $(1/[A]_{0,i}, 1/v_{0,i})$ using our definition of **v₀[a_]** from the previous example.

```
lbData =
```

$$\text{Table}\left[\left\{\text{recipa}, \frac{1}{\text{v}_0\,[1\,/\,\text{recipa}]}\right\}, \{\text{recipa}, 1, 5001, 1000\}\right];$$

We then use **ListPlot** to graph this **Table** as well as **Plot** to graph the inverse of v_0 [a_] directly.

```
gph1 = ListPlot[lbData,
    PlotStyle -> {PointSize[0.02]}, DisplayFunction → Identity];
gph2 = Plot[─────────────, {recipa, -2001, 5001},
             v₀[1 / recipa]
    DisplayFunction → Identity];

Show[gph1, gph2, DisplayFunction → $DisplayFunction,
    AxesLabel -> {"1/[A]", "1/v₀"}];
```

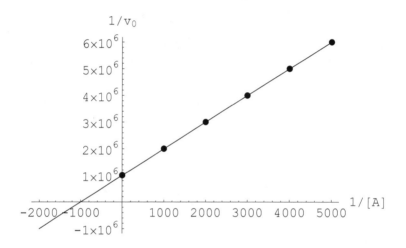

Figure 2.2. Lineweaver-Burk plot for a simple Michaelis-Menten enzyme reaction. Note the ordinate and abscissa intercepts occur at $1/V_{max}$ and $-1/K_m$, respectively.

2.1.5 Eadie-Hofstee plot

The Eadie-Hofstee equation is derived by multiplying both sides of Eqn [2.1] by $(K_m + [A]_0)$, dividing by $[A]_0$, and then rearranging the terms to give:

$$v_0 = -K_m \frac{v_0}{[A]_0} + V_{max} \ . \tag{2.3}$$

Hence, a plot of the data pairs consisting of $(v_{0,i}/[A]_{0,i}, v_{0,i})$ gives a straight line with a slope that has the value $-K_m$ and an ordinate intercept that is the value of V_{max}.

Q: Generate an Eadie-Hofstee plot for the enzyme described in the worked example in Section 2.1.2.

A: First we define the Eadie-Hofstee equation

```
v₀[vOnA0_] := -Kₘ vOOnA0 + Vₘₐₓ ;
```

and then we can generate the appropriate **Table** of ordered pairs and graph as follows:

```
ehData =
  Table[{vOOnA0, v₀[vOOnA0]}, {vOOnA0, 0, 0.001, 0.0001}];

gph1 = ListPlot[ehData, PlotStyle -> {PointSize[0.025]},
    DisplayFunction → Identity];
gph2 = Plot[v₀[vOOnA0], {vOOnA0, 0, 0.001},
    DisplayFunction → Identity];

Show[gph1, gph2, DisplayFunction → $DisplayFunction,
    AxesLabel -> {"v₀/[A]", "v₀"}];
```

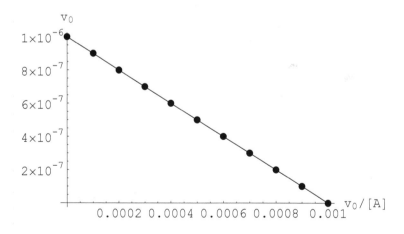

Figure 2.3. Eadie-Hofstee plot for a simple Michaelis-Menten enzyme reaction. Note that the slope gives the value of $-K_m$ and the ordinate intercept gives the value of V_{max}.

2.1.6 Hanes-Woolf plot

The Hanes-Woolf equation is derived by taking the reciprocal of both sides of Eqn [2.1] and multiplying each by $[A]_0$ and then rearranging terms to give:

$$\frac{[A]_0}{v_0} = \frac{[A]_0}{V_{max}} + \frac{K_m}{V_{max}} \ . \tag{2.4}$$

Thus, for a Michaelis-Menten enzyme reaction, a plot of the data pairs ($[A]_{0,i}$, $[A]_{0,i}/v_{0,i}$) gives a straight line with a slope whose value is $1/V_{max}$ and an abscissa intercept that is $-K_m$.

2.1.7 Eisenthal and Cornish-Bowden equation (direct linear plot)

A totally different approach to analyzing enzyme kinetic data involves plotting many lines onto the experimental data; it was introduced in 1974 by Eisenthal and Cornish-Bowden.[3] For this procedure a Cartesian axis-system is drawn with v_0 on the ordinate and $[A]_0$ on the abscissa. With the data pairs ($[A]_{0,i}$, $v_{0,i}$), $i = 1,...,N$, a straight line is drawn to pass through the points ($-[A]_{0,i}$, 0) and (0, $v_{0,i}$). The intersection of this line with others, similarly drawn, occurs at as many different points as there are data pairs. In other words, since experimental data are never error free, a set of values $\{K_m, V_{max}\}_j$, $j = 1,...,N-1$, is obtained. The next step in the analysis is to arrange the respective values of $K_{m,j}$ and $V_{max,j}$ into two separate sets, in ascending order. It has been shown that the 'statistically best estimate' of the two parameters is given by the respective median (middle) values in the set.[3]

2.2 Enzyme Inhibition

2.2.1 Degree of inhibition

The rate of an enzymic reaction is often affected by substances other than the reactant(s). *Inhibitors* slow the rate and *activators* increase it. In dealing with inhibitors it is important to distinguish between the effects that are observed experimentally on the various analytical plots mentioned above, and the molecular mechanism (or models) proposed to explain the effects. The purpose behind studies of enzyme inhibition is often to enhance an understanding of the mechanism of the enzyme by interpreting the changes in apparent values of V_{max} and K_m in terms of various possible models of the mechanism (see Chapter 3).

There are three basic types of inhibition of an enzyme and these are defined in terms of the *degree of inhibition,* which is defined as

$$i = \frac{v_0 - v_0^i}{v_0} \ , \tag{2.5}$$

where v_0 and v_0^i are the initial velocities of the reaction measured in the absence and presence of the inhibitor, respectively.

1. *Pure noncompetitive* inhibition is said to exist if i is unaffected by the concentration of the substrate.

2. *Competitive* inhibition exists if i decreases as the substrate concentration is increased.

3. *Anti-* or *uncompetitive* inhibitions exists if i increases as the substrate concentration is increased.

In addition to these canonical forms of inhibition is *mixed* inhibition, in which i increases or decreases as the substrate concentration increases, but not to the same extent as for the pure competitive or anticompetitive cases. Mechanistically, noncompetitive inhibition is a special case of mixed inhibition, but *operationally,* as defined in Eqn [2.5], mixed inhibition is a combination of two of the above canonical types of inhibition.

2.2.2 Michaelis-Menten equations that include inhibitor effects

There are four simple equations[4] that are extensions of the Michaelis-Menten one shown as Eqn [2.1]. These equations merit special consideration because the kinetics of many enzymes can be satisfactorily described by them. In these equations K_I and K_I' are inhibition constants that are equilibrium dissociation constants and as such have units of $mol\ L^{-1}$.

1. Pure noncompetitive inhibition : $\quad v_0 = \dfrac{V_{max}[A]_0}{(K_m + [A]_0)\left(1 + \frac{[I]}{K_I}\right)}$. $\qquad$ [2.6]

2. Pure competitive inhibition : $\quad v_0 = \dfrac{V_{max}[A]_0}{K_m\left(1 + \frac{[I]}{K_I}\right) + [A]_0}$. $\qquad$ [2.7]

3. Anticompetitive inhibition : $\quad v_0 = \dfrac{V_{max}[A]_0}{K_m + [A]_0\left(1 + \frac{[I]}{K_I}\right)}$. $\qquad$ [2.8]

4. Mixed inhibition : $\quad v_0 = \dfrac{V_{max}[A]_0}{K_m\left(1 + \frac{[I]}{K_I}\right) + [A]_0\left(1 + \frac{[I]}{K_I'}\right)}$. $\qquad$ [2.9]

2.3 Enzyme Mechanisms

2.3.1 Michaelis-Menten mechanism

Michaelis and Menten set out to explain the mechanistic basis of the peculiar kinetic data that they had obtained from the invertase-catalyzed hydrolysis of sucrose. This reaction yields glucose and fructose from sucrose. In particular they wanted to understand the general relationship between the rate of the reaction catalyzed by a given amount of enzyme in a given volume and the concentration of sucrose. It had already been shown by others that the relationship was not linear and that the reaction rate increased to a maximum value, and not beyond, as the substrate concentration was increased. This outcome exemplified the phenomenon known as *saturation*. Their mechanistic interpretation of the data involved postulating that the enzyme forms a complex with sucrose and the idea that this complex then undergoes hydrolysis at a rate that is slow relative to the rate of formation of the enzyme-sucrose complex. The reaction scheme they proposed was

$$\text{E} + \text{A} \underset{k_{-1}}{\overset{k_1}{\rightleftharpoons}} \text{EA} \overset{k_2}{\longrightarrow} \text{E} + \text{P} \ . \tag{2.10}$$

Thus, Michaelis and Menten assumed that the values of both k_1 and k_{-1} are very large compared with that of k_2, and the *overall* rate of the reaction depends on the concentration of the EA complex, so the reaction rate can be expressed as

$$v_0 = k_2 [\text{EA}] \ . \tag{2.11}$$

To derive a formula for v_0, an expression is needed for [EA], but this can only be given in terms of parameters that are experimentally *knowable*. These include the fact that the total concentration of the enzyme and the reactants are each constant; hence, so-called *conservation of mass* conditions are written as

$$[\text{A}]_0 = [\text{A}] + [\text{EA}] + [\text{P}] \ , \tag{2.12}$$

$$[\text{E}]_0 = [\text{E}] + [\text{EA}] \ . \tag{2.13}$$

Since it is assumed that $k_1, k_{-1} \gg k_2$, the first part of the reaction can be described as if it is in a state of *quasi*-equilibrium, so the relationship between the concentrations of the reactants is described by an equilibrium constant, $K_A = [\text{E}] [\text{A}]/[\text{EA}]$. The expression for the equilibrium constant can be rearranged to give one for the concentration of the EA complex in terms of the knowables in Eqns [2.12] and [2.13]:

$$K_A = ([\text{E}]_0 - [\text{EA}]) [\text{A}]_0 / [\text{EA}] \ , \tag{2.14}$$

which, upon rearrangement, gives

$$[\text{EA}] = \frac{[\text{E}]_0 [\text{A}]_0}{K_A + [\text{A}]_0} \ . \tag{2.15}$$

Substitution of Eqn [2.15] into Eqn [2.11] gives

$$v_0 = \frac{k_2 [\text{E}]_0 [\text{A}]_0}{K_A + [\text{A}]_0} \ . \tag{2.16}$$

This equation is of exactly the same form as Eqn [2.1] provided that we equate $V_{\max}$ with $k_2 [\text{E}]_0$, and K_m with K_A.

The choice of the particular reaction scheme in Eqn [2.10], and the assumption about the relative rates of the two sub-reactions in the overall scheme, lead to the hyperbolic equation that describes the dependence of the rate of the enzyme-catalyzed reaction on the substrate concentration. Thus, a macroscopically observable (rectangular hyperbola) response to an experimental variable was able to be interpreted in terms of an *invisible* molecular process. Hence this is a *classical* example of what science is all about!

2.3.2 The steady state

In 1925 Briggs and Haldane[5] reinvestigated the theory of Michaelis and Menten and introduced an important new development. Instead of assuming that the first stage of the reaction (Eqn [2.14]) is in equilibrium, they assumed that the *rate of change* of the concentration of the EA complex is (approximately) zero, i.e., it is in a steady state. This situation is described mathematically as

$$\frac{d[EA]}{dt} = 0 \ .$$ [2.17]

Then the flux equation for [EA] is given by

$$\frac{d[EA]}{dt} = k_1 [E] [A] - (k_{-1} + k_2) [EA] \ ,$$ [2.18]

and by using the conservation of mass conditions (Eqns [2.12] and [2.13]) with $[A] \simeq [A]_0$, then

$$k_1 [E]_0 [A]_0 - (k_1 [A]_0 + k_{-1} + k_2)[EA] = 0 \ .$$ [2.19]

After rearranging this equation and using the substitution that $v_0 = k_2 [EA]$, the result is

$$v_0 = \frac{k_2 [E]_0 [A]_0}{\frac{k_{-1} + k_2}{k_1} + [A]_0} \ .$$ [2.20]

Remarkably, the equation has the same form as Eqns [2.1] and [2.16], provided that K_m is identified with $(k_{-1} + k_2)/k_1$ and V_{max} has the same interpretation as before, namely, $k_2 [E]_0$. Thus, Briggs and Haldane provided a new mechanistic interpretation of the rectangular hyperbolic relationship between v_0 and $[A]_0$. In summary, this was based on the idea of an EA complex and the fact that, to a good level of approximation because the rate of formation of the complex is fast relative to its breakdown to free enzyme and product, its time derivative is effectively zero, or it is said to be in a steady state.

There are several features of Eqn [2.20] that are worth noting:

1. Because k_2 describes the number of molecules of substrate converted to product per second, per molecule of enzyme, it is called the *turnover number* of the enzyme. In general, if the enzyme has a mechanism that is more complex than that in Eqn [2.10] the expression for V_{max} is more complicated than simply $k_2 [E]_0$, and it is an expression that is the ratio of sums of products of unitary rate constants (see later in Chapter 3 for examples).

2. If the enzyme is not pure it may not be possible to know accurately the concentration of the active form of the enzyme, $[E]_0$. Nevertheless, V_{max} can still be obtained by using steady-state kinetic analysis. Thus, in order to standardize experimental results,

the enzyme activity is usually expressed in enzyme units, which are the amount of enzyme that transforms 1 µmol of substrate into product(s) in 1 min, under standard conditions of pH, ionic strength, and temperature (usually 25°C).

3. When $[A]_0$ is large compared with K_m, virtually all of the active enzyme is in the form of the EA complex, so the enzyme is said to be saturated with substrate. In this situation the enzyme operates at its maximum velocity under the specified conditions of buffer-type, pH, temperature, etc.

2.3.3 Reversible Michaelis-Menten enzyme

A more realistic representation of the reaction scheme for an enzyme, as opposed to Eqn [2.10], is one in which the conversion of the EA complex to product is reversible. Thus the scheme is

$$E + A \underset{k_{-1}}{\overset{k_1}{\rightleftharpoons}} EA \underset{k_{-2}}{\overset{k_2}{\rightleftharpoons}} E + P \ . \tag{2.21}$$

The derivation of the steady-state rate equation for this mechanism is begun with the general expression for the overall, or net, rate of product formation:

$$v = \frac{d[P]}{dt} = k_2[EA] - k_{-2}[E][P] \ . \tag{2.22}$$

Hence, the rate equations for [EA] and [E] are required. They are as follows:

$$\frac{d[EA]}{dt} = k_1[E][A] - (k_{-1} + k_2)[EA] + k_{-2}[E][P] \ , \tag{2.23}$$

and

$$\frac{d[E]}{dt} = -k_1[E][A] + (k_{-1} + k_2)[EA] - k_{-2}[E][P] \ . \tag{2.24}$$

The steady-state assumption specifies that

$$\frac{d[EA]}{dt} = 0 \ , \tag{2.25}$$

and

$$\frac{d[E]}{dt} = 0 \ . \tag{2.26}$$

Thus, by combining Eqns [2.23] and [2.25] with the conservation of mass conditions (Eqn [2.12] and [2.13]), and the fact that the substrate and product concentrations are much greater than total enzyme concentration, we obtain

$$k_1([E]_0 - [EA])[A]_0 - (k_{-1} + k_2)[EA] + k_{-2}([E]_0 - [EA])[P]_0 = 0 \ , \tag{2.27}$$

so

$$(k_1[A]_0 - k_{-2}[P]_0)[E]_0 + (k_1[A]_0 + k_{-1} + k_2 + k_{-2}[P]_0) = 0 \ , \tag{2.28}$$

and

$$[EA] = \frac{k_1 [A]_0 - k_{-2} [P]_0}{k_1 [A]_0 + k_{-1} + k_2 + k_{-2} [P]_0} \quad . \tag{2.29}$$

Similarly, the expression for [E] is obtained by combining Eqns [2.24] and [2.26] with the conservation of mass conditions, to give

$$[E] = \frac{(k_1 + k_{-2})[E]_0}{k_1 [A]_0 + k_{-1} + k_2 + k_{-2} [P]_0} \quad . \tag{2.30}$$

Hence, from Eqns [2.22] and [2.30] and after some algebraic simplification, the rate expression becomes:

$$v_0 = \frac{(k_1 k_2[A]_0 - k_{-1} k_{-2} [P]_0)[E]_0}{k_1 [A]_0 + k_{-1} + k_2 + k_{-2} [P]_0} \quad . \tag{2.31}$$

Q: At first sight, Eqn [2.31] has a different form from that of the Michaelis-Menten expression (Eqn [2.1]); however, it becomes the same under certain specific conditions. To reveal this carry out the following: Make [P] = 0 and then algebraically rearrange the terms in Eqn [2.31].

A: Define Eqn [2.31] with

```
Clear[Subscript] ;
```

$$v_0[a\_, p\_] := \frac{(k_1 \ k_2 \ a - k_{-1} \ k_{-2} \ p) \ e_0}{k_1 \ a + k_{-1} + k_2 + k_{-2} \ p} \ ;$$

Note that the definition of a function of two variables is a simple extension of the one variable case. By setting $p \to 0$ we obtain

```
vel = v₀[a, p] /. p -> 0
```

$$\frac{a \ e_0 \ k_1 \ k_2}{k_{-1} + a \ k_1 + k_2}$$

This expression has the same form as the Michaelis-Menten equation; it is more clearly seen by the following algebraic re-arrangements:

```
num = Numerator[vel] / k₁ ;
denom = FactorTerms[Denominator[vel] / k₁, a ];
```

$$\frac{num}{denom} \ /. \ e_0 \ k_2 \ -> V_{max} \ /. \ \frac{k_{-1}}{k_1} + \frac{k_2}{k_1} \ -> K_m$$

$$\frac{a \ V_{max}}{a + K_3}$$

Numerator[*expr*]	numerator of *expr*
Denominator[*expr*]	denominator of *expr*
FactorTerms[*expr*, *x*]	pull out factors that do not depend on *x*

Rearranging algebraic expressions.

Q: The result in the previous example is the Michaelis-Menten expression. What happens to Eqn [2.31] when k_{-2} is set equal to zero, thus making the reaction scheme in Eqn [2.21] the same as that in Eqn [2.10]?

A: Use the following replacement procedure.

vel = v₀[a, p] /. k₋₂ -> 0

$$\frac{a\ e_0\ k_1\ k_2}{k_{-1} + a\ k_1 + k_2}$$

Additional algebraic re-arrangements lead to the next expression.

num = Numerator[vel] / k₁ ;
denom = FactorTerms[Denominator[vel] / k₁, a] ;

$$\frac{num}{denom}\ /.\ e_0\ k_2\ ->\ V_{max}\ /.\ \frac{k_{-1}}{k_1} + \frac{k_2}{k_1}\ ->\ K_m$$

$$\frac{a\ V_{max}}{a + K_3}$$

Thus, we can see that this again is the Michaelis-Menten expression.

2.4 Regulatory Enzymes

The enzyme kinetic equations discussed so far can be called Michaelis-Menten ones. A plot of reaction velocity versus substrate concentration is a rectangular hyperbola. Alternatively, a Lineweaver-Burk plot is a straight line. This situation pertains not only to many single-substrate enzymes but it also applies to many enzymes with two or more substrates when the reaction velocity is measured as a function of the concentration of one of the substrates while the others are held constant (see Chapter 3).

On the other hand, many other enzymes display non-linear Lineweaver-Burk plots, or plots of velocity versus substrate concentration that are not simple rectangular hyperbolas. Such plots are described by mathematical functions that are a ratio of polynomials, in which the numerator is one degree less than that of the denominator. The enzyme mechanism that can be analyzed to yield the rate equation usually involves an oligomeric enzyme with allosterically regulated active sites. For examples of rate equations commonly used to model such enzymes, see the excellent books by Cornish-Bowden[6] and Roberts[7] on enzyme kinetics.

2.5 Exercises

2.5.1

Explore the functional behavior of the *Michaelis-Menten* equation (Eqn [2.1]) by (1) altering the value of the maximum substrate concentration. (2) Notice the extent to which V_{max} is likely to be able to be accurately inferred from the graph. (3) What happens to the shape of the plot when K_m is varied?

2.5.2

Investigate the functional behavior of the Lineweaver-Burk equation (Eqn [2.2]) by doing the following. (1) Alter the value of the maximum substrate concentration and notice the extent to which V_{max} is likely to be able to be accurately inferred from the graph. (2) What happens to the abscissa intercept when the value of K_m is decreased?

2.5.3

Investigate the functional behavior of the Eadie-Hofstee equation (Eqn [2.3]) by doing the following. (1) Alter the value of V_{max} and evaluate what happens to the ordinate intercept. (2) What happens to the abscissal intercept when K_m is decreased?

2.5.4

Plot the Hanes-Woolf equation (Eqn [2.4]) and explore its functional behaviour by doing the following. (1) Alter the value of V_{max} and evaluate what happens to the slope. (2) What happens to the value of the abscissa intercept when K_m is decreased?

2.6 References

1. Sumner, J.B. (1926) Isolation and crystallization of urease. *J. Biol. Chem.* **69**, 433-441.
2. Michaelis, L. and Menten, M.L. (1913) Die kinetik der invertinwirkung. *Biochem Z.* **49**, 333-369.
3. Eisenthal, R. and Cornish-Bowden, A. (1974) Direct linear plot. New graphical procedure for estimating enzyme kinetic parameters. *Biochem. J.* **139**, 715-720.
4. Kuchel, P.W. and Ralston, G.B. (Eds.) (1998) *Schaum's Outline of Theory and Problems of Biochemistry.* 2nd ed. McGraw-Hill, New York, pp. 251-289.
5. Briggs, G.E. and Haldane, J.B.S. (1925) Note on the kinetics of enzyme action. *Biochem. J.* **19**, 338-339.
6. Cornish-Bowden, A. (1995) *Fundamentals of Enzyme Kinetics.* Portland Press, London.
7. Roberts, D.V. (1977) *Enzyme Kinetics.* Cambridge University Press, New York.

3 Basic Procedures for Simulating Metabolic Systems

3.1 Introduction

Traditionally the study of enzyme kinetics has been divided into two main areas, steady-state analysis and pre-steady-state analysis. The former usually requires simple instrumentation like a spectrophotometer with which to perform the assays, but the latter involves special apparatus that is fitted to a spectrophotometer to record the progress of a reaction in its first few milli- or even microseconds. Practitioners of pre-steady-state kinetics use stopped-flow and temperature-jump devices and they were often held in awe by the 'mere mortals' who used more traditional methods. In recent years other techniques including radioactive-isotope exchange and NMR magnetization transfer have added to the approaches for gaining data to help define the mechanisms of enzymic reactions; these methods enable the measurement of fluxes in the forward and reverse directions when the overall reaction is at chemical equilibrium and they give access to the fast steps in a reaction. What makes NMR even more significant is that it can be applied to intact cells and tissues and it has even been used to study fast membrane-transport reactions that occur in the sub-second time scale.

Often, the sets of kinetic parameter values available in the literature on a given enzyme are diverse, having been obtained with many different experimental techniques and under a range of experimental conditions. It is sometimes difficult to reach a satisfactory conclusion regarding which values in a set of kinetic parameters to use in a computer model that involves the particular enzyme. In addition, the sets of parameters are usually incomplete and are insufficient to define unambiguously the underlying reaction mechanism. Therefore, a systematic approach is needed to arrive at a set of parameter values, preferably unitary rate constants, that are consistent with the proposed mechanism of the enzyme and fit in with the known values of the steady-state parameters.

Thus, there are three main aims in this chapter. They are to (1) show how to take a set of steady-state kinetic parameters for an enzyme and generate a set of unitary rate constants that are internally consistent with this set; (2) incorporate into a reaction mechanism features that reflect the pH dependence of the enzyme; and (3) account for the effect of cation binding to the enzyme and its substrate(s) in a mechanistic and kinetic way.

3.1.1 Inborn errors of metabolism

Experimentally, the most readily available kinetic data on enzymes are those obtained with steady-state analysis. In particular, in the literature the most abundant quantitative kinetic items are the values of K_m and V_{max}. Inherited systematic changes in these parameter values in a particular enzyme for each particular disease is the basis of inborn errors of metabolism. Part of the motivation for this book is the expectation that a quantitative understanding of the metabolic changes, based on simulations like those described herein, will lead to more rational treatments.

Inborn errors of metabolism come about through several fundamentally different mechanisms that can ultimately be traced backed to mutations in genes. Alterations of the base sequence of a gene can lead to (1) altered packaging of DNA which changes the rate of transcription to messenger RNA; (2) altered processing of mRNA including splicing changes and hence altered amino acid sequences in the protein; (3) altered rate of translation of the processed mRNA into protein; (4) altered amino acid sequence in the progenitor enzyme or the final enzyme molecule. Thus the inherited defect can lead to (5) a reduction in the amount of the otherwise normal enzyme; (6) defective targeting of the enzyme to intracellular or plasma membrane locations that expresses itself as a reduction in total effective enzyme concentration and hence V_{max}; (7) substitution of an amino acid residue, or even its omission, leads to altered binding affinity of the enzyme for its substrate(s); this changes K_m values; (8) a similar outcome occurs for an inhibitor or effector binding site which affects the value of K_I or K_E; (9) alteration of the interface between domains of a monomeric enzyme or the subunits of an oligomeric or polymeric enzyme and hence a change in the 'interaction constant' L (e.g., Section 2.5).

3.1.2 Constancy of K_{eq}

All of the possible structural bases of altered enzymic activity are ultimately reflected in the values of the unitary rate constants, and hence in the corresponding steady-state parameters, of the enzyme. Thus, when modelling inborn errors of metabolism it is usual to focus on a particular step in the reaction mechanism and to change the values of the two or more relevant unitary rate constants. It is worth remembering that an enzyme does not change the value of the equilibrium constant for the overall (bio)chemical reaction; it only changes the rate at which equilibrium is attained. For example, consider modelling a reduced binding affinity of an enzyme for one of its substrates. At least one forward rate constant will need to be decreased in order to reflect the reduced rate, but to preserve the constancy of the equilibrium constant for the whole reaction, at least one of the 'off' or reverse rate constants must also be reduced.

3.2 Relationships between Unitary Rate Constants and Steady-State Parameters

3.2.1 Progress curve of a Michaelis-Menten reaction

Consider the irreversible Michaelis-Menten scheme – it is the simplest enzyme mechanism of all (see also Eqn [2.31]).

$$E + A \underset{k_{-1}}{\overset{k_1}{\rightleftharpoons}} EA \overset{k_2}{\longrightarrow} E + P \quad , \tag{3.1}$$

$$v_0 = \frac{k_2 [E]_0 [A]_0}{\left(\frac{k_{-1} + k_2}{k_1}\right) + [A]_0} \quad . \tag{3.2}$$

Eqn [3.2] expresses the fact that in a conventional steady-state kinetic analysis of such an enzyme only two parameters are estimated, K_m and V_{max}, and these encapsulate all the information that is required to describe the progress of the reaction after the establishment of a steady state of the EA concentration. In other words, if we assume that [EA] is constant, which involves ignoring the rapid pre-steady-state phase of the time course, the progress of the reaction is described by the solution of the Michaelis-Menten differential equation.

Q: Simulate the decline of substrate in a reaction that is catalyzed by a simple Michaelis-Menten enzyme; let the K_m and V_{max} be 5 mmol L^{-1} and 100.0 μmol L^{-1} min^{-1}, respectively, and let the initial concentration of the substrate be 10 mmol L^{-1}. Simulate a time course of 400 min.

A: The relevant differential equation is (Eqn [2.1])

$$\frac{d[A]}{dt} = -\frac{V_{max}[A]_0}{K_m + [A]_0} \quad , \tag{3.3}$$

and a suitable *Mathematica* program entails the use of **NDSolve** (see Section 1.4.6), as follows:

```
Clear[Subscript];
```

$$v_1 := \frac{V_{max}\, a[t]}{K_m + a[t]} ;$$

```
Vmax = 100 10^-6;
Km = 5.0 10^-3;
a0 = 10.0 10^-3;

solution =
  NDSolve[{a'[t] == -v₁,  a[0] == a0 }, a[t], {t, 0, 400}]

{{a[t] → InterpolatingFunction[{{0., 400.}}, <>][t]}}
```

By using the resulting **InterpolatingFunction** (Section 1.4.6) we can **Plot** the time course.

```
Plot[Evaluate[a[t] /. solution], {t, 0, 400},
  AxesLabel -> {"Time (min)", "Concentration (M)"}];
```

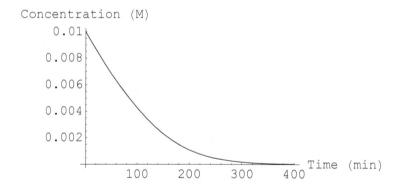

Figure 3.1. Time course of the enzymic reaction described by Eqn [3.3].

Q: In the previous example we only simulated the decline of A. How do we simulate the appearance of P?

A: The conservation of mass condition for A specifies that

$$[A]_0 = [A]_t + [P]_t \quad , \tag{3.4}$$

where the subscript t denoted any given time, hence

$$[P]_t = [A]_0 - [A]_t \ . \tag{3.5}$$

So, using our **solution** from the previous example, we must **Plot** Eqn [3.5] to solve this problem.

```
Plot[Evaluate[{a[t], a0 - a[t]} /. solution], {t, 0, 400},
    AxesLabel -> {"Time (min)", "Concentration (M)"}];
```

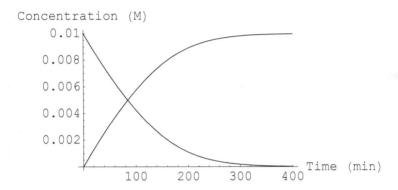

Figure 3.2. Time course of substrate and product concentration of the Michaelis-Menten reaction.

We note that the two progress curves are mirror images of each other with the horizontal line of reflection being at half the starting concentration of A. This outcome is a direct consequence of the conservation relationship in Eqn [3.4]. For the Michaelis-Menten reaction scheme this is not strictly true since there will be some buildup of the enzyme-substrate complex EA (see Eqn [2.12]) prior to it attaining a steady state. But as long as [A] is much greater than the concentration of the enzyme, this will usually be an exceptionally good approximation for times beyond a second or less.

3.2.2 Pre-steady-state Michaelis-Menten scheme

Eqn [3.2] contains the definition of the steady-state kinetic parameters of a simple Michaelis-Menten enzyme:

$$K_m = \frac{k_{-1} + k_2}{k_1} \quad , \tag{3.6}$$

and

$$V_{max} = k_2 [E]_0 \ , \tag{3.7}$$

where $[E]_0$ is the molar concentration of the active sites of the enzyme.

If we wish to simulate the pre-steady-state phase of the reaction, it is necessary to specify the values of the unitary rate constants. In the absence of any experimental studies that give these values, or at least some of them, we must solve the nonlinear algebraic equations that are expressed in Eqns [3.6] and [3.7] for the three values $(k_1, k_{-1}, \text{and } k_2)$.

It is usually possible to make a reliable estimate of the turnover number of an enzyme (Section 2.3.2) from an estimate of V_{max}, hence k_2 is readily evaluated by rearranging Eqn [3.7]. This outcome is possible provided that the enzyme has been purified and its molecular weight, and order of oligomerization (see Section 3.2.3), are known. Solving Eqns [3.6] and [3.7] for the three unknowns clearly means that the value of one of the unitary rate constants must be arbitrarily specified. Suppose we give k_1 the value aa $mol^{-1} \ L \ s^{-1}$, then symbolic expressions for the other two unitary rate constants can be obtained by using the **Solve** function.

Solve[*lhs* == *rhs*, *vars*]   attempts to solve an equation or
                                      set of equations for the variables *vars*

Solving simultaneous equations.

Q: What is the procedure for determining expressions for the unitary rate constants for a Michaelis-Menten enzyme in terms of the steady-state parameters V_{max} and K_m?

A: This problem, which is one of solving a pair of simultaneous nonlinear algebraic equations, is solved as follows:

```
Clear[Subscript];

k₁ = aa;
          k₋₁ + k₂
eqn₁ := Kₘ == ─────────── ;
             k₁
eqn₂ := Vmax == k₂ e₀;

Solve[{eqn₁, eqn₂}, {k₋₁, k₂}]
```

$$\left\{ \left\{ k_{-1} \rightarrow -\frac{-aa \ e_0 \ K_m + V_{max}}{e_0}, \ k_2 \rightarrow \frac{V_{max}}{e_0} \right\} \right\}$$

The analysis for this problem is rather elementary and the equations could easily have been done accurately without symbolic computation. However, the basic approach is applicable to sets of much more complex expressions of steady-state parameters so it is a strategy well worth understanding.

3.2.3 Enzyme oligomerization and the turnover number

There is a complication with the definition of $[E]_0$; it pertains to the concentration of active sites since many enzymes are oligomeric. If the enzyme is oligomeric then $[E]_0$ must be related to the concentration of the holoenzyme (or whole oligomer), $[E']_0$, by the expression

$$[E]_0 = n \times m \times [E']_0 \ ,$$ [3.8]

where n is the number of subunits in the oligomeric holoenzyme and m is the number of active sites on each monomer (subunit). If we know the concentration of the holoenzyme, the value of the forward catalytic breakdown rate constant is related to the turnover number of each active site by the expression

$$k_{cat}^f = turnover \ number \times n \times m \ ,$$ [3.9]

where m is almost invariably 1, and the maximum velocity is given by

$$V_{max} = k_{cat}^f [E']_0 \ .$$ [3.10]

3.2.4 Specific examples of enzyme mechanisms

Consider an example of the above analysis for a simple hydrolytic enzyme, arginase. Arginase catalyses the hydrolysis of arginine to ornithine and urea.

$$arginine \rightarrow ornithine + urea \ .$$ [3.11]

Although we will return to this enzyme a little later, in this instance let us ignore the fact that there are two products of the reaction and assume that the kinetics of arginase are well described by the Michaelis-Menten equation. Then it is simple to relate the unitary rate constants to the steady-state kinetic parameters.

Q: Determine a set of unitary rate constants that are consistent with the known steady-state parameters for a 1 ng L^{-1} solution of arginase assuming it to be a simple Michaelis-Menten enzyme with a K_m for arginine of 5.0 mmol L^{-1} and a turnover number of 4.5 $\times 10^3$ s^{-1}. The molecular mass of arginase is ~105,000 Da and it exists in solution as a trimer.

A: First, we use Eqn [3.8] to convert the gram concentration to a molar concentration of enzyme.

$$\mathbf{edash_0} = 1. \times 10^{-9} \times 10^3 \; / \, 105000$$

$$9.52381 \times 10^{-12}$$

By applying Eqn [3.8] we obtain the concentration of subunits in the trimer.

$$\mathbf{e_0} = 3 \times 1 \times \mathbf{edash_0}$$

$$2.85714 \times 10^{-11}$$

Therefore, by using Eqns [3.6] and [3.7] we obtain the following values for the Michaelis-Menten parameters:

$$\mathbf{V_{max}} = \mathbf{e_0} \times 4.5 \times 10^3 \; ;$$
$$\mathbf{K_m} = 5.0 \times 10^{-3} ;$$

Finally, by arbitrarily selecting a value of $1 \times 10^7 \, \text{mol}^{-1} \, \text{L} \, \text{s}^{-1}$ for k_1, we can use **Solve** with Eqns [3.6] and [3.7] to determine the values of k_{-1} and k_2.

$$\mathbf{k_1} = 1.0 \times 10^7 ;$$

$$\mathbf{eqn_1} := \mathbf{K_m} == \frac{\mathbf{k_{-1}} + \mathbf{k_2}}{\mathbf{k_1}} ;$$

$$\mathbf{eqn_2} := \mathbf{V_{max}} == \mathbf{k_2} \, \mathbf{e_0} ;$$

$$\mathbf{solution} = \mathbf{Solve}[\{\mathbf{eqn_1}, \mathbf{eqn_2}\}, \{\mathbf{k_{-1}}, \mathbf{k_2}\}]$$

$$\{\{\mathbf{k_{-1}} \to 45500., \mathbf{k_2} \to 4500.\}\}$$

The answer to the question is that the unitary rate constants that are consistent with the values of the steady-state parameters are $\{k_1 = 1 \times 10^7 \, \text{mol}^{-1} \, \text{L} \, \text{s}^{-1}, k_{-1} = 4.55 \times 10^4$ $\text{s}^{-1}, k_2 = 4.5 \times 10^3 \, \text{s}^{-1}\}$.

Q: Use the unitary rate constants determined in the previous example, in conjunction with the relevant differential equations, to simulate a time course of the arginase-catalyzed reaction. Compare the results with those obtained by solving the time course using the Michaelis-Menten equation (Eqn [3.3]).

A: With the parameters and solution still defined from the last example and assuming that the initial concentration of arginine is 10 mM, we solve the time course by using the unitary rate constants and rate equations with the following input:

$$\mathbf{a0} = 10.0 \times 10^{-3} ;$$

$$\mathbf{preSSTimecourse} = \mathbf{NDSolve}[\{$$
$$\mathbf{a'[t]} == -\mathbf{k_1} \, \mathbf{a[t]} \, \mathbf{e[t]} + \mathbf{k_{-1}} \, \mathbf{ea[t]} ,$$
$$\mathbf{ea'[t]} == \mathbf{k_1} \, \mathbf{a[t]} \, \mathbf{e[t]} - (\mathbf{k_{-1}} + \mathbf{k_2}) \, \mathbf{ea[t]} ,$$

```
e'[t]  == -k₁ a[t] e[t] + (k₋₁ + k₂) ea[t],
p'[t]  == k₂ ea[t],
a[0] == a0, ea[0] == 0, e[0] == e₀, p[0] == 0.0} /. solution,
{a, ea, e, p},
{t, 0, 1}, AccuracyGoal -> 12];
```

Note the application of the **/. solution** operator to the list of differential equations. This introduces the values of k_{-1} and k_2 that were obtained in the previous example; k_1 is still an assigned value. We have also specified the option **AccuracyGoal →** **12** (see Section 1.6.4). The reason for this is that we are interested in the concentration of the product in the 'pre-steady-state' phase of the reaction that occurs within the first few milliseconds when the concentration of product is very low.

Similarly, we solve for the time course using the Michaelis-Menten rate equation (Eqn [3.3]).

```
ssTimecourse =
```

$$\text{NDSolve}\left[\left\{a'[t] == -\frac{V_{max}\, a[t]}{K_m + a[t]},\quad a[0] == a0\right\},\right.$$
$$\left.a[t],\ \{t,\ 0,\ 1\},\ \text{AccuracyGoal} \to 12\right];$$

By graphing the two time courses we obtain

```
Plot[
   {Evaluate[p[t] /. preSSTimecourse],
    Evaluate[(a0 - a[t]) /. ssTimecourse]},
   {t, 0.0, 1},
   AxesLabel → {"Time (min)", "Concentration (M)"},
   PlotRange -> All];
```

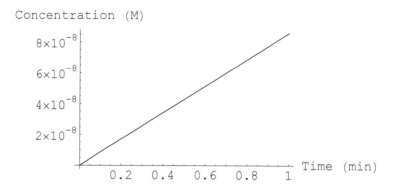

Figure 3.3. Time course of the arginase-catalyzed reaction. Comparing pre-steady-state reaction kinetics to Michaelis-Menten kinetics.

It is evident that under the conditions used and in the timescale considered, there is no perceptible difference between the pre-steady-state and steady-state solutions for the time course. However, if we plot just the first 100 μs of the timecourse,

```
Plot[
    {Evaluate[p[t] /. preSSTimecourse],
     Evaluate[(a0 - a[t]) /. ssTimecourse]},
    {t, 0.0, 0.0001},
    AxesLabel → {"", "Concentration (M)"}, PlotRange -> All];
```

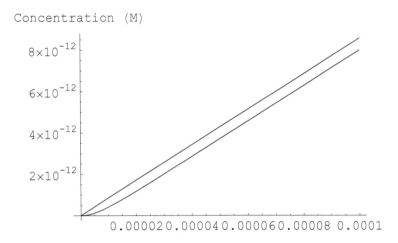

Figure 3.4. Time course of the arginase-catalyzed reaction. Comparing pre-steady-state reaction kinetics (lower curve) to Michaelis-Menten kinetics (upper curve) within the first 100 μs.

Thus, we see a small time lag before the pre-steady-state time course has the same slope as that described by the solution of the steady-state equation.

3.3 Upper Limit of Values for Unitary Rate Constants

In the first question in Section 3.2.4 we sought the value of three unitary rate constants. However, there were only two steady-state parameters so it was necessary to assume the value of one of either k_1 or k_{-1}; k_1 was chosen because there is a rational choice that can be made for the value of such a second-order rate constant. Could we have chosen a much larger value than the $1 \times 10^7 \, \text{mol}^{-1} \, \text{L s}^{-1}$ that was used? The answer is "yes, sort of!" as there is an *upper bound* on the value; this is called the 'diffusion limit' for a second-order rate constant.

3.3.1 Diffusion control of reaction rate

The upper limit of second-order rate constants can be determined experimentally by using rapid-reaction methods. At ~37°C the value has been found to be ~1×10^8 $\text{mol}^{-1} \, \text{L s}^{-1}$. This value applies to solutions such as those found in the cytoplasm of a cell, or in the plasma of blood. A simple theoretical analysis shows how this value comes about.

Consider two reacting molecular species, A and B, in solution. If the rate of the reaction is *proportional* to the rate at which a molecule of A collides with one of B, then the effective mean intermolecular (inter-collision) distance and the diffusion coefficient of each reactant will dictate the rate. Since the molecules are in motion we can, without loss of generality, conveniently focus attention on just one molecule of A and place its centre at the origin of a spherical polar coordinate system. The relative motion of the two molecules can then be characterized by the sum of their diffusion coefficients ($D_A + D_B$). If the inward flux of molecules of B through a sphere of radius r around the single molecule of A is denoted by J (flux; units, mol m^{-2} s^{-1}), then the magnitude of the flux is given by Fick's first law of bulk diffusion,

$$J = -(D_A + D_B) \frac{d[n_B]}{dr} \quad , \tag{3.12}$$

where the derivative term expresses the gradient in number-concentration (number of molecules per unit volume) of B. Thus, the number of molecules of B diffusing across the surface of a sphere of radius r (surface area $= 4 \pi r^2$) per second is given by $I = J 4 \pi r^2$. Equation [3.12] can be integrated with respect to the radius and [n_B] in the bulk medium,

$$\frac{I}{4 \pi r^2} \int_{r=R_{AB}}^{r=R_\infty} \frac{dr}{r^2} = -(D_A + D_B) \int_0^{[n_B] = [n_B]_{r=\infty}} d[n_B] \quad , \tag{3.13}$$

where $([n_B])_{r=\infty}$ is defined as the concentration in the bulk solution, namely, $[n_B]_{bulk}$. Thus,

$$\frac{-I}{4 \pi R_\infty} + \frac{I}{4 \pi R_{AB}} = (D_A + D_B) ([n_B])_{bulk} \quad , \tag{3.14}$$

and at $r = R_\infty$ the flux is assumed to be zero. Eqn [3.14] is rearranged to give (dropping the "bulk" subscript) the expression that has units of (number of molecules)$^{-1}$ m^3 s^{-1},

$$\frac{I \text{ (number of molecules s}^{-1})}{[n_B] \text{ (number of molecules m}^{-3})} = k' = 4 \pi R_{AB} (D_A + D_B) \quad ; \tag{3.15}$$

[n_B] is now expressed as a molar concentration by dividing Eqn [3.15] by Avogadro's number, N, and multiplying by 10^{-3} to convert the units of m^{-3} to L^{-1}. Hence, multiplying the Eqn [3.15] by $N \times 10^3$ gives the expression for the second-order rate constant, k, with the appropriate units (mol^{-1} L s^{-1}),

$$k = k' N 10^3 = 4 \pi R_{AB} (D_A + D_B) N 10^3 \quad . \tag{3.16}$$

Q: Does the analysis in the previous section (Eqn [3.16]) yield a realistic value for the upper limit of a second-order rate constant that characterizes the reaction between a metabolite and enzyme in a cellular environment?

A: The diffusion coefficient of bulk water at 37°C is ~2 × 10^{-9} m^2 s^{-1}. Suppose that the metabolite B has a diffusion coefficient that is an order of magnitude less than water because of the higher viscosity in the cell, and the enzyme's diffusion coefficient is smaller again by a factor of 10 because of its much larger molecular mass. Also, the radius of many metabolites is ~ 0.2 to 1 nm while that of a typical enzyme is ~3 nM. By using Eqn [3.16] the value of k is calculated to be

$$k = 4\pi * 4 \times 10^{-9} \, m \, (2.2 \times 10^{-10} \, m^2 \, s^{-1}) \, 6.022 \, 10^{23} \, mol^{-1} \, 10^3 \, L \, m^{-3}$$

$$\frac{6.65937 \times 10^9 \, L}{mol \, s}$$

This value is about two orders of magnitude greater that we expected if the claim of 1×10^8 mol^{-1} L s^{-1} is correct. The explanation for the discrepancy lies in the very simplistic model used in our analysis. We did not take into account the fact that molecules need to be correctly aligned in order to react, so not all collisions lead to products. Furthermore, the value of the diffusion coefficients under cellular conditions are likely to be significantly less than that used here. So, 1×10^8 mol^{-1} L s^{-1} appears to be a useful, albeit conservatively low, value to use in simulating enzymic systems. This value is useful if we are faced with having to choose an arbitrary one for a second-order rate constant in simulations of metabolism, and it is used in some of the following sections.

3.4 Realistic Enzyme Models

In reality, it is uncommon to encounter enzymic reactions that can be simply described by the Michaelis-Menten mechanism. Those enzymes that appear most commonly to conform to this simple model are hydrolytic ones. As this name suggests water is a co-reactant with the substrate, so the enzyme has in reality two substrates and not one. Because the concentration of water in a reaction mixture is so large (55.5_5 mol L^{-1}), relative to the other substrate(s), it is assumed to be constant and therefore it is not usual to include the concentration of water in the rate equation. On the other hand, it is not possible to ignore the fact that two products are formed in most hydrolytic reactions. A good example of this is again provided by arginase, an enzyme of the urea cycle that we first considered in Section 3.2.4.

3.4.1 Deriving steady-state rate equations

The steady-state rate equation for any enzymic reaction mechanism can, in principle, be derived in the same way as that for the Michaelis-Menten mechanism (Section 2.3.2).

Q: Derive an expression for the steady-state rate equation for arginase in terms of its unitary rate constants.

A: The complete reaction mechanism of arginase is

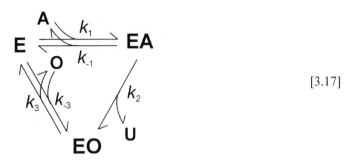

[3.17]

where A, U, and O denote arginine, urea, and ornithine, respectively.

A rate equation for this mechanism can be derived by using the steady-state assumption described in Section 2.3.2. First we write expressions that describe the rate of change of the concentrations of all but one enzyme form.

```
Clear[k, Subscript];

eqn₁ := e'[t] == -k₁ e a + k₋₁ ea - k₋₃ e o + k₃ eo;
eqn₂ := ea'[t] == k₁ e a - k₋₁ ea - k₂ ea;
```

By applying the steady-state assumption, the left-hand side of each equation is set equal to zero.

```
e'[t] = 0;
ea'[t] = 0;
```

Expressions describing the steady-state concentrations of each of the enzyme forms can then be derived by solving the above simultaneous equations and the conservation of mass equation for the enzyme.

```
solution = Solve[{eqn₁, eqn₂, e0 == e + ea + eo}, {e, ea, eo}]
```

$$\left\{\left\{ eo \to -\frac{-e0\, o\, k_{-3}\, k_{-1} - e0\, o\, k_{-3}\, k_2 - a\, e0\, k_1\, k_2}{o\, k_{-3}\, k_{-1} + o\, k_{-3}\, k_2 + a\, k_1\, k_2 + k_{-1}\, k_3 + a\, k_1\, k_3 + k_2\, k_3},\right.\right.$$

$$e \to \frac{e0\,(k_{-1} + k_2)\, k_3}{-a\, k_1\,(k_{-1} - k_3) - (k_{-1} + k_2)\,(-o\, k_{-3} - a\, k_1 - k_3)},$$

$$\left.\left. ea \to \frac{a\, e0\, k_1\, k_3}{-a\, k_1\,(k_{-1} - k_3) - (k_{-1} + k_2)\,(-o\, k_{-3} - a\, k_1 - k_3)}\right\}\right\}$$

The rate of production of urea is given by

```
varg := u'[t] == k₂ ea
```

and hence the steady-state rate equation is

```
varg /. solution
```

$$\{u'[t] == \frac{a\,e0\,k_1\,k_2\,k_3}{-a\,k_1\,(k_{-1} - k_3) - (k_{-1} + k_2)\,(-o\,k_{-3} - a\,k_1 - k_3)}\}$$

From the last question, the steady-state rate equation for arginase is

$$v = \frac{d[U]}{dt} = \frac{k_1\,k_2\,k_3\,[A]\,[E]_0}{k_3(k_{-1} + k_2) + k_{-3}\,(k_{-1} + k_2)\,[O] + k_1\,(k_2 + k_3)\,[A]}, \qquad [3.18]$$

and the correspondences between the unitary rate constants and the steady-state parameters are readily derived.

$$k_{cat}^{f} = V_{max}^{f}\,/\,[E]_0 = k_2\,k_3\,/\,(k_2 + k_3), \qquad [3.19]$$

$$K_{m,A} = k_3\,(k_{-1} + k_2)\,/\,(k_1\,(k_2 + k_3)), \qquad [3.20]$$

$$K_{i,O} = k_3\,/\,k_{-3}. \qquad [3.21]$$

Note the use of nomenclature to denote the forward catalytic steady-state rate constant, k_{cat}^{f}, and its corresponding maximum velocity. Also, since there are Michaelis and inhibition constants for each reactant, it is necessary to use double subscripts such as $K_{m,A}$ and $K_{i,O}$, respectively.

3.4.2 The `RateEquation` function

Another way to derive steady-state rate equations is to use the **RateEquation** function that is fully described in Appendix 1. It automatically derives a steady-state rate equation for almost any enzyme mechanism with up to 14 enzyme forms. Before this function is used for the first time you must evaluate all the cells in Appendix 1 so that the appropriate .m file is created. This allows the function to be called using the following commands.

`<< rateequationderiver`	loads in the function **RateEquation**
`RateEquation[rcm, el]`	derives the steady state rate equation for an enzyme mechanism defined in the rate constant matrix (*rcm* – see below and Appendix 1). The argument *el* is optional and is a list of user defined names for the enzyme forms of the reaction mechansim.

Deriving rate equations.

Q: Use the function **RateEquation** to derive the rate expression given in Eqn [3.18].

A: First it is necessary to load in the function **RateEquation**.

<< rateequationderiver

Then, the only 'thinking task' required of you is to set up the rate constant matrix based on the reaction scheme in Eqn [3.17]. The matrix is

$$
\begin{array}{cccc}
 & \rightarrow \text{e} & \rightarrow \text{ea} & \rightarrow \text{eo} \\
\text{e} \rightarrow & 0 & k_1\,a & k_{-3}\,o \\
\text{ea} \rightarrow & k_{-1} & 0 & k_2 \\
\text{eo} \rightarrow & k_3 & 0 & 0
\end{array}
$$

Hence, in *Mathematica* the matrix is entered as

rcmArginase = {{0, k$_1$ a, k$_{-3}$ o}, {k$_{-1}$, 0, k$_2$}, {k$_3$, 0, 0}};

The function **RateEquation** is applied as follows:

RateEquation[rcmArginase, {e, ea, eo}]

The output is

```
Enzyme Distribution Functions
```

```
e/eo =  (k₋₁ k₃ + k₂ k₃)
```
$$e/eo = (k_{-1}\,k_3 + k_2\,k_3)$$

```
ea/eo =  a k₁ k₃
```
$$ea/eo = a\,k_1\,k_3$$

```
eo/eo =  (o k₋₃ k₋₁ + o k₋₃ k₂ + a k₁ k₂)
```
$$eo/eo = (o\,k_{-3}\,k_{-1} + o\,k_{-3}\,k_2 + a\,k_1\,k_2)$$

```
Steady-State Rate Equations
```

```
d[a]/dt =-a eo k₁ k₂ k₃/ Denominator
```
$$d[a]/dt = -a\,eo\,k_1\,k_2\,k_3 / \text{Denominator}$$

```
d[o]/dt =a eo k₁ k₂ k₃/ Denominator
```
$$d[o]/dt = a\,eo\,k_1\,k_2\,k_3 / \text{Denominator}$$

```
Denominator
```

```
o (k₋₃ k₋₁ + k₋₃ k₂) + k₋₁ k₃ + k₂ k₃ + a (k₁ k₂ + k₁ k₃)
```
$$o\,(k_{-3}\,k_{-1} + k_{-3}\,k_2) + k_{-1}\,k_3 + k_2\,k_3 + a\,(k_1\,k_2 + k_1\,k_3)$$

Thus, it can be seen that the functional form of the output is the same as Eqn [3.18], as required by the question.

3.4.3 Calculating a consistent set of unitary rate constants

Having determined the steady-state rate equation, as well as the relationships between the unitary rate constants and the steady-state kinetic parameters, it is possible to calculate a consistent set of unitary rate constants.

Q: Calculate a consistent set of unitary rate constants for the arginase mechanism shown in Eqn [3.17] by using the three well-known[8] values of the steady-state parameters for the enzyme, $K_{m,A}$, $K_{i,O}$ and the turnover number.

A: Use the *Mathematica* function **Solve** as follows:

The known parameter values are[8]

```
kcat,f = 4.5 × 10³; (*mol L⁻¹ s⁻¹*)
Km,a = 5.0 × 10⁻³; (*mol L⁻¹ *)
Ki,o = 3.0 × 10⁻³; (*mol L⁻¹ *)
```

Only three steady-state parameter values are known, so two of the five unitary rate constants must be assigned values. Select the two second-order rate constants and set them to 1/10th of the diffusion limit value (Section 3.3.1).

```
k₁ = 1.0 × 10⁷;
k₋₃ = 1.0 × 10⁷;
```

Now use **Solve** to determine the values of the unitary rate constants from the non-linear algebraic equations that relate the steady-state parameters to the unitary rate constants (Section 3.4.1).

```
Solve[{kcat,f == k₂ k₃ / (k₂ + k₃),
    Km,a == k₃ (k₋₁ + k₂) / (k₁ (k₂ + k₃)), Ki,o == k₃ / k₋₃}, {k₋₁, k₂, k₃}]

{{k₋₁ → 53529.4, k₂ → 5294.12, k₃ → 30000.}}
```

Thus the set of unitary rate constants whose values are consistent with the values of the steady-state parameters are $\{k_1 = 1 \times 10^7 \text{ mol}^{-1} \text{ L s}^{-1}, k_{-1} = 5.35 \times 10^4 \text{ s}^{-1}, k_2 = 5.29 \times 10^3 \text{ s}^{-1}, k_3 = 3 \times 10^4 \text{ s}^{-1}, k_{-3} = 1 \times 10^7 \text{ mol}^{-1} \text{ L s}^{-1}\}$.

3.5 Deriving Expressions for Steady-State Parameters

Enzymes are often studied by using a range of concentrations of various substrates, products, and inhibitors in order to determine the reaction mechanism. The choice of particular experiments is made on the basis of the well-trodden path of standard enzyme kinetic practice, as described in many authoritative texts.[4–7]

As noted in Section 3.4.2, the function **RateEquation** provides a means of rapidly deriving the steady-state rate equations for virtually any enzyme mechanism. But for the present problem we must regroup the various terms involving unitary rate constants in the rate equation to couch the rate equation on a form that has the equivalent Michaelis-Menten constants, inhibition constants, and V_{max} parameters. For arginase (Section 3.4.1) the relationships between the unitary rate constants and the steady-state parameters were readily apparent. However, for many enzyme mechanisms this is not the case.

The V_{max} of an enzymic reaction is simply the rate obtained when the substrate concentration is (infinitely!) high. This situation ensures that the active site(s) of the enzyme are saturated with the substrate. When choosing a range of substrate concentrations for use in kinetic experiments the rule of thumb that is used is that "the highest substrate concentration must be at least 5 times the K_m value" for the particular substrate. Recall that the K_m is defined as the substrate concentration that gives half the maximum velocity. Then at the top of this range the enzyme will be operating at approximately 5/6 th of its V_{max}.

For enzymes with two or more co-substrates, V_{max} is the velocity when each co-substrate is at a saturating concentration, and the products are all at zero concentration. The corresponding K_m values are now defined as the substrate concentration that gives the half-maximum velocity when all the other co-substrates are at saturating concentrations.

The biochemical literature is replete with kinetic data on enzymes with the maximal velocities often expressed in terms of the *turnover number(s)*, and the steady-state parameters given as Michaelis constants and various types of inhibition constant. The challenge facing the metabolic modeler is to match these values with the appropriate terms in the rate equation that describes the appropriate enzyme mechanism. Hence, it is useful to have an automatic procedure for deriving the expressions for these parameters. This process is shown in the next example by using another urea cycle enzyme, ornithine carbamoyl transferase.

Q: Derive the steady-state rate equation and hence expressions for V_{max} in the forward and reverse directions, as well as the respective K_m values for each substrate, in the ornithine carbamoyl transferase reaction.

A: The reaction mechanism for this urea cycle enzyme is[8]

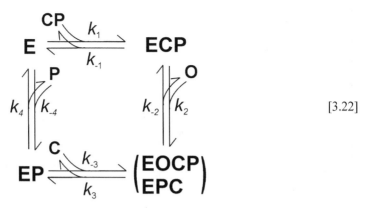

$$[3.22]$$

where CP, O, C, and P denote carbamoyl phosphate, ornithine, citrulline, and orthophosphate, respectively. There are two important points to note about this mechanism. First, the formation of a ternary complex occurs between the enzyme and both substrates; and second, the substrates and products bind/release to/from the enzyme in a specific order. These two features lead to the definition of this mechanism as a *compulsory-ordered ternary-complex* one.

We derive the overall steady-state expression for the reaction mechanism of Eqn [3.22] in a manner similar to that used in the example in Section 3.4.2. We begin by setting up the rate constant matrix and then using **RateEquation**, as follows:

```
Clear[Subscript];

rcmOCT = {{0, k₁ cp, 0, k₋₄ p},
     {k₋₁, 0, k₂ o, 0}, {0, k₋₂, 0, k₃}, {k₄, 0, k₋₃ c, 0}};

RateEquation[rcmOCT, {e, ecp, eocp, ep}]
```

The output is

Enzyme Distribution Functions

$e/eo = (c\ k_{-3}\ k_{-2}\ k_{-1} + k_{-2}\ k_{-1}\ k_4 + k_{-1}\ k_3\ k_4 + o\ k_2\ k_3\ k_4)$

$ecp/eo = (c\ p\ k_{-4}\ k_{-3}\ k_{-2} + c\ cp\ k_{-3}\ k_{-2}\ k_1 + cp\ k_{-2}\ k_1\ k_4 + cp\ k_1\ k_3\ k_4)$

$eocp/eo = (c\ p\ k_{-4}\ k_{-3}\ k_{-1} + c\ o\ p\ k_{-4}\ k_{-3}\ k_2 + c\ cp\ o\ k_{-3}\ k_1\ k_2 + cp\ o\ k_1\ k_2\ k_4)$

$ep/eo = (p\ k_{-4}\ k_{-2}\ k_{-1} + p\ k_{-4}\ k_{-1}\ k_3 + o\ p\ k_{-4}\ k_2\ k_3 + cp\ o\ k_1\ k_2\ k_3)$

```
Steady-State Rate Equations
```

```
d[cp]/dt =eo (c p k₋₄ k₋₃ k₋₂ k₋₁ - cp o k₁ k₂ k₃ k₄)/ Denominator
```

```
d[p]/dt =eo (-c p k₋₄ k₋₃ k₋₂ k₋₁ + cp o k₁ k₂ k₃ k₄)/ Denominator
```

```
d[o]/dt =eo (c p k₋₄ k₋₃ k₋₂ k₋₁ - cp o k₁ k₂ k₃ k₄)/ Denominator
```

```
d[c]/dt =eo (-c p k₋₄ k₋₃ k₋₂ k₋₁ + cp o k₁ k₂ k₃ k₄)/ Denominator
```

```
Denominator
```

```
c k₋₃ k₋₂ k₋₁ + p (k₋₄ k₋₂ k₋₁ + c (k₋₄ k₋₃ k₋₂ + k₋₄ k₋₃ k₋₁) +
      k₋₄ k₋₁ k₃ + o (c k₋₄ k₋₃ k₂ + k₋₄ k₂ k₃)) +
   k₋₂ k₋₁ k₄ + k₋₁ k₃ k₄ + o k₂ k₃ k₄ + cp
      (c k₋₃ k₋₂ k₁ + k₋₂ k₁ k₄ + k₁ k₃ k₄ + o (c k₋₃ k₁ k₂ + k₁ k₂ k₃ + k₁ k₂ k₄))
```

We can now apply the **Limit** function in various ways to determine the steady-state parameters.

Limit[*expr*, $x \to x_0$]   finds the limiting value of *expr* when x approaches x_0

The Limit function.

We start by choosing one of the rate expressions in the output generated above; it is usual practice to select an expression that describes the rate of change of concentration of a product (since we usually define the rate of product formation to be positive). We convert this particular expression to an input one by cutting and pasting it into an input Cell:

vel = (-c p k₋₃ k₋₃ k₋₂ k₋₁ + cp o k₁ k₂ k₃ k₄) / denom;

denom =
    c k₋₃ k₋₂ k₋₁ + p (k₋₄ k₋₂ k₋₁ + c (k₋₄ k₋₃ k₋₂ + k₋₄ k₋₃ k₋₁) + k₋₄ k₋₁ k₃ +
       o (c k₋₄ k₋₃ k₂ + k₋₄ k₂ k₃)) + k₋₂ k₋₁ k₄ +
    k₋₁ k₃ k₄ + o k₂ k₃ k₄ + cp (c k₋₃ k₋₂ k₁ + k₋₂ k₁ k₄ +
       k₁ k₃ k₄ + o (c k₋₃ k₁ k₂ + k₁ k₂ k₃ + k₁ k₂ k₄));

Now derive the expression for V_{max} for the forward direction. This entails taking the limit of the rate equation as the two substrate concentrations go to infinity while also setting the product concentrations to zero.

```
Vmaxf1 = Limit[vel, o → Infinity];
Vmaxf2 = Limit[Vmaxf1, cp → Infinity];
Vmax,f = Vmaxf2 /. c → 0;
Print["Vmax in forward direction = ", Vmax,f]
```

$$\text{Vmax in forward direction} = \frac{k_3\ k_4}{k_3 + k_4}$$

The expression for the K_m of ornithine is obtained by (1) solving the rate equation for the value of the ornithine concentration that yields a rate of $V_{max}/2$; (2) setting the product concentrations to zero; and (3) taking the limit as the co-substrate(s) go to infinity. The analysis uses the powerful symbolic capability of *Mathematica* for this otherwise very tedious and error-prone task.

```
kmorn1 = Solve[vel == Vmax,f / 2 , o];
kmorn2 = kmorn1 /. {c → 0, p → 0};
Km,o = o /. Limit[kmorn2, cp → Infinity];
Print["Km for ornithine = ", Km,o];
```

$$\text{Km for ornithine} = \left\{ \frac{(k_{-2} + k_3)\ k_4}{k_2\ (k_3 + k_4)} \right\}$$

Similarly, we obtain expressions for the other steady-state parameters. For example, V_{max} for the reverse direction is obtained in the same manner as for the forward direction but with the products now viewed as substrates.

```
Vmaxr1 = Limit[vel, c → Infinity];
Vmaxr2 = Limit[Vmaxr1, p → Infinity];
Vmax,r = Vmaxr2 /. o → 0;
Print["Vmax  in reverse direction =", Vmax,r];
```

$$\text{Vmax in reverse direction} = -\frac{k_{-2}\ k_{-1}}{k_{-2} + k_{-1}}$$

The other K_m expressions are determined as follows:

```
(*Km for carbamoyl phosphate is...*)
kmcp1 = Solve[vel == Vmax,f / 2 , cp];
kmcp2 = kmcp1 /. {c → 0, p → 0};
Km,cp = cp /. Limit[kmcp2, o → Infinity];
```

```
Print["Km for carbamoyl phosphate = ", Km,cp];

(*Km for citrulline is...*)
               Vmax,r
kmcit1 = Solve[vel == ──────, c];
                  2
kmcit2 = kmcit1 /. {o → 0, cp → 0};
Km,c = c /. Limit[kmcit2, p → Infinity];
Print["Km for citrulline = ", Km,c];

(*Km for phosphate is...*)
               Vmax,r
kmpi1 = Solve[vel == ──────, p];
                  2
kmpi2 = kmpi1 /. {o → 0, cp → 0};
Km,p = p /. Limit[kmpi2, c → Infinity];
Print["Km for phosphate = ", Km,p];
```

$$\text{Km for carbamoyl phosphate} = \left\{ \frac{k_3\, k_4}{k_1\, (k_3 + k_4)} \right\}$$

$$\text{Km for citrulline} = \left\{ \frac{k_{-1}\, (k_{-2} + k_3)}{k_{-3}\, (k_{-2} + k_{-1})} \right\}$$

$$\text{Km for phosphate} = \left\{ \frac{k_{-2}\, k_{-1}}{k_{-4}\, (k_{-2} + k_{-1})} \right\}$$

Having derived the relationships between the unitary rate constants and the experimentally measurable steady-state parameters, a consistent set of unitary rate constants can be calculated (see next worked example).

Q: Calculate a consistent set of unitary rate constants for the ornithine carbamoyl transferase mechanism shown in Eqn [3.22]. Use the three known values of the steady-state parameters and the equilibrium constant.[8]

A: In this mechanism there are eight unitary rate constants but only three well-known steady-state parameter values, plus the equilibrium constant of the overall reaction. Therefore it is necessary to assume values of four of the unitary constants. As in the example in Section 3.4.3, the program is also based on the **Solve** function.

The known parameter values are[8]

$$k_{cat,f} = 1.4 \times 10^3;$$
$$K_{m,cp} = 8.1 \times 10^{-5};$$
$$K_{m,o} = 9.0 \times 10^{-4};$$
$$K_{eq} = 1.0 \times 10^5;$$

and we assume the following 'reasonable' values for four of the unitary rate constants:

$$k_{-2} = 1.0 \times 10^3;$$
$$k_3 = 3.0 \times 10^3;$$
$$k_{-3} = 9.0 \times 10^4;$$
$$k_{-4} = 5.0 \times 10^5;$$

We now **Solve** for the unitary rate constants:

$$\texttt{Solve}\Big[\Big\{k_{cat,f} == \frac{k_3\,k_4}{k_3 + k_4}\,,\ K_{m,cp} == \frac{k_3\,k_4}{k_1\,(k_3 + k_4)}\,,\ K_{m,o} == \frac{(k_{-2} + k_3)\,k_4}{k_2\,(k_3 + k_4)}\,,$$
$$K_{eq} == k_1\,k_2\,k_3\,k_4\,/\,(k_{-1}\,k_{-2}\,k_{-3}\,k_{-4})\Big\},\ \{k_1,\ k_{-1},\ k_2,\ k_4\}\Big]$$

$$\{\{k_{-1} \to 62.7343,\ k_1 \to 1.7284 \times 10^7,\ k_2 \to 2.07407 \times 10^6,\ k_4 \to 2625.\}\}$$

In conclusion, the unitary rate constants that are consistent with the values of the steady-state parameters are $\{k_1 = 1.73 \times 10^7\ \text{mol}^{-1}\ \text{L s}^{-1},\ k_{-1} = 62.73\ \text{s}^{-1},\ k_2 = 2.07 \times 10^6\,\text{mol}^{-1}\ \text{L s}^{-1},\ k_3 = 3 \times 10^3\ \text{s}^{-1},\ k_{-3} = 9 \times 10^4\ \text{mol}^{-1}\ \text{L s}^{-1},\ k_4 = 2.63 \times 10^3\ \text{s}^{-1}$, and $k_{-4} = 2.07 \times 10^6\ \text{mol}^{-1}\ \text{L s}^{-1}\}$. These are used in a model of the urea cycle that is described in Section 3.8.

3.6 Multiple Equilibria

Another major task in formulating a model of metabolism is the inclusion of quantitative descriptions of the concentrations of complexes between metal cations and, primarily, phosphorylated metabolic intermediates. The procedure used to calculate these concentrations turns out to be formally equivalent to that used to determine the values of unitary rate constants, given a set of steady-state kinetic parameters. In the present case the initial concentrations of all reactants must be specified, together with expressions for, and values of, the various equilibrium constants of the binding reactions. Having made this claim it is probably most convincing to simply illustrate the process of analysis; it is as follows.

Q: Calculate the concentrations of free Mg, free ATP, and their 1:1 complex in a reaction mixture that has attained equilibrium, from a total concentration of 3 mmol L^{-1} Mg and 1 mmol L^{-1} ATP. Assume that the reaction scheme is

$$\text{Mg}^{2+} + \text{ATP}^{4-} \underset{}{\overset{K_{\text{MgATP}}}{\rightleftharpoons}} \text{MgATP}^{2-}, \quad K_{\text{MgATP}} = 1 \times 10^2\ \text{M}^{-1} \quad . \qquad [3.23]$$

A: We begin by writing the expression for the equilibrium constant of the reaction; it is

$$eqn_1 := K_{mgatp} == \frac{mgatp2mi}{mg \times atp4mi};$$

$$K_{mgatp} = 1 \times 10^2;$$

The following conservation of mass conditions also apply.

$$eqn_2 := mg + mgatp2mi == 3.0 \times 10^{-3};$$

$$eqn_3 := atp4mi + mgatp2mi == 1.0 \times 10^{-3};$$

Thus there exist three equations in three unknowns which we solve using the function **Solve**.

```
solution = Solve[{eqn₁, eqn₂, eqn₃}, {mg, atp4mi, mgatp2mi}]
```

```
{{mgatp2mi → 0.00021767, mg → 0.00278233, atp4mi → 0.00078233},
 {mgatp2mi → 0.0137823, mg → -0.0107823, atp4mi → -0.0127823}}
```

The output is a list of two solutions but we note by inspection that only one of them is physically meaningful because all concentrations must be positive. The only meaningful solution shows that the concentration of free Mg is 2.782 mM, and that of its 1:1 complex with ATP is 0.218 mM. A simple way to delete the nonphysical solution is with the following input which relies on a number of *Mathematica* pattern recognition commands.

```
realSolution = solution /.
     ({a___, b_ -> c_, d___} /; Negative[c]) -> {} // Flatten
```

```
{mgatp2mi → 0.00021767, mg → 0.00278233, atp4mi → 0.00078233}
```

It is left to the reader to use the *Mathematica* help browser to understand these commands and syntax of this input.

Also by way of a check on the solution to the problem we note that the conservation of mass condition for total ATP is satisfied since 2.782 mM + 0.218 mM = 3.0 mM. The test for conservation of mass is always worth carrying out to ensure your program is functioning correctly. This can be tested with the following input:

```
{eqn₂, eqn₃} /. realSolution
```

```
{True, True}
```

({a___, b_ -> c_, d___} /; Negative[c]) -> {}	replacement rule which converts a list of rules containing negative values on the *rhs* to an empty set
list // **Flatten**	removes the inner set of braces from the *list*

Removing non-physically meaningful solutions from a list of replacement rules.

Q: Consider the more realistic reaction scheme than that given above in which ATP exists in various protonated forms and these have different values of the binding constant for Mg and Ca. In addition, suppose that Ca is present in the solution as well. Calculate the concentrations of free Mg, Ca, and the various protonated forms of ATP and their complexes in a reaction mixture that has attained equilibrium from a total concentration of each species of 3 mmol L^{-1}. Assume that the reaction scheme is as follows, noting that here we use the superscripted valences of the ions to help emphasize the net charges of the different complexes.

The complexes of Mg and ATP are formed in the following reactions with their respective equilibrium constants.

$$Mg^{2+} + ATP^{4-} \xrightleftharpoons{K_{MgATP^{2-}}} MgATP^{2-}, \quad K_{MgATP^{2-}} = 1 \times 10^2 \, M^{-1} \quad , \qquad [3.24]$$

$$Mg^{2+} + ATP^{3-} \xrightleftharpoons{K_{MgATP^-}} MgATP^-, \quad K_{MgHATP^-} = 7 \times 10^3 \, M^{-1} \quad . \qquad [3.25]$$

Complexes of Ca and ATP are

$$Ca^{2+} + ATP^{4-} \xrightleftharpoons{K_{CaATP^{2-}}} CaATP^{2-}, \quad K_{CaATP^{2-}} = 1 \times 10^2 \, M^{-1} \quad , \qquad [3.26]$$

$$Ca^{2+} + ATP^{3-} \xrightleftharpoons{K_{CaATP^-}} CaATP^-, \quad K_{CaATP^-} = 1 \times 10^4 \, M^{-1} \quad . \qquad [3.27]$$

Protonation of ATP also occurs.

$$H^+ + ATP^{4-} \xrightleftharpoons{K_{ATP^{3-}}} ATP^{3-}, \quad K_{ATP^{3-}} = 3 \times 10^6 \, M^{-1} \quad , \qquad [3.28]$$

$$H^+ + ATP^{3-} \xrightleftharpoons{K_{ATP^{2-}}} ATP^{2-}, \quad K_{ATP^{2-}} = 1 \times 10^4 \, M^{-1} \quad . \qquad [3.29]$$

A: We solve this problem in a manner similar to that used in the previous question, by setting up the expressions for the equilibrium constants and defining the conservation of mass conditions.

The equilibrium equations for the reactions in Eqns [3.24 - 3.29] are

$$\text{eqn}_1 := K_{mgatp2mi} == \frac{mgatp2mi}{mg \times atp4mi} \; ; \qquad\qquad K_{mgatp2mi} = 1 \times 10^2 \; ;$$

$$\text{eqn}_2 := K_{mgatpmi} == \frac{mgatpmi}{mg \times atp3mi} \; ; \qquad\qquad K_{mgatpmi} = 7 \times 10^3 \; ;$$

$$\text{eqn}_3 := K_{caatp2mi} == \frac{caatp2mi}{ca \times atp4mi} \; ; \qquad\qquad K_{caatp2mi} = 1 \times 10^2 \; ;$$

$$\text{eqn}_4 := K_{caatpmi} == \frac{caatpmi}{ca \times atp3mi} \; ; \qquad\qquad K_{caatpmi} = 1 \times 10^4 \; ;$$

$$eqn_5 := K_{atp3mi} \times h == \frac{atp3mi}{atp4mi} \; ; \qquad\qquad K_{atp3mi} = 3 \times 10^6 \; ;$$

$$eqn_6 := K_{atp2mi} \times h == \frac{atp2mi}{atp3mi} \; ; \qquad\qquad K_{atp2mi} = 1 \times 10^4 \; ;$$

The conservation of mass equations are

```
eqn₇ := mg + mgatp2mi + mgatpmi == 3.0 × 10⁻³;
eqn₈ := ca + caatp2mi + caatpmi == 3.0 × 10⁻³;
eqn₉ := atp4mi + atp3mi + atp2mi +
        caatp2mi + caatpmi + mgatp2mi + mgatpmi == 5.0 × 10⁻³;
```

Hence there are 9 equations and 10 unknowns. By setting the pH to 7.2 and then calculating the H^+ concentration, we can solve for the remaining 9 unknowns.

```
pH = 7.2;
h = 1 * 10^-pH;

unknowns = {mg, ca, atp4mi, atp3mi,
    atp2mi, mgatp2mi, mgatpmi, caatp2mi, caatpmi};

solution = Solve[{eqn₁, eqn₂, eqn₃,
    eqn₄, eqn₅, eqn₆, eqn₇, eqn₈, eqn₉}, unknowns];

realsolution = solution /.
    ({a___, b_ -> c_, d___} /; Negative[c]) -> {} // Flatten;

MatrixForm[Transpose[{unknowns, unknowns /. realsolution}]]
```

$$
\begin{pmatrix}
mg & 0.00122309 \\
ca & 0.000989526 \\
atp4mi & 0.00101951 \\
atp3mi & 0.00019298 \\
atp2mi & 1.21762 \times 10^{-7} \\
mgatp2mi & 0.000124695 \\
mgatpmi & 0.00165222 \\
caatp2mi & 0.000100883 \\
caatpmi & 0.00190959
\end{pmatrix}
$$

Note that we have used the new commands **MatrixForm** and **Transpose** to display the solutions as a matrix.

> **MatrixForm[*list*]**   diplays the list in matrix form
>
> **Transpose[*m*]**   Transposes the matrix *m*

Displaying lists.

Papers by Conigrave and Morris[1] and Mulquiney and Kuchel[2,3] treat even more complicated reaction schemes than those that are considered here, but the basic principles of solving the nonlinear algebraic equations to yield estimates of all the concentrations of the reactants are exactly the same. The nonlinearity of the system comes about via the definitions of the equilibrium constants, since the expressions are ratios of products of the equilibrium concentrations of the reactants. The other important aspect of setting up the analysis is the definition of the conservation of mass conditions. In these expressions special care is needed when taking into account the stoichiometry of the complexes.

3.7 pH Effects on Kinetic Parameters

The dependence of enzymic activity on pH is well known and the simplest reaction schemes that encapsulate this dependence are elaborations of the Michaelis-Menten one. In general, changes come about in V_{max} and K_m because of changes in one or a combination of the following: (1) ionization of groups in the substrate; (2) ionization of groups involved in binding the substrate; (3) groups involved in catalysis; and (4) ionization of other groups on the enzyme such as effector binding sites.

Because the topic is well described in many books on enzyme kinetics,[4-7] only one illustrative example of a pH effect on a simple enzyme will be given here. This is the simplest of all enzymic-reaction schemes.

3.7.1 Ionization of the substrate

If we assume that there are no relevant ionizable groups on E or EA, then the scheme that describes just ionization of the substrate is

$$\begin{array}{c} A \\ \updownarrow Ka \\ HA^+ \end{array} + \; E \; \underset{k_{-1}}{\overset{k_1}{\rightleftharpoons}} \; EA \; \overset{k_2}{\longrightarrow} \; E + P \; . \qquad [3.30]$$

This is analogous to pure competitive inhibition (Section 2.2.2) and hence the rate equation is

$$v = \frac{V_{max}[A]_0}{K_m(1 + [H^+]/K_a) + [A]_0} \; , \qquad [3.31]$$

where K_a denotes the *acid* (hence the subscript *a*) dissociation constant of the HA^+ complex and it is given by

$$K_a = \frac{[H^+]_e [A]_e}{[HA^+]} \quad .$$

[3.32]

Note that at the start of the reaction

$$[A]_0 = [A] + [HA^+] \quad .$$

[3.33]

It is also evident that the apparent Michaelis constant, K_m', depends on pH, with the protons acting like a competitive inhibitor.

$$K_m' = K_m\left(1 + \frac{[H^+]}{K_a}\right) \quad .$$

[3.34]

Thus, if the pH is decreased the apparent Michaelis constant will increase because there is a decrease in the concentration of the 'true' substrate, HA^+.

Another model of substrate ionization entails the enzyme binding the protonated form of the substrate, then the rate equation is the same as Eqn [3.31] except that

$$K_m\left(1 + \frac{[H^+]}{K_a}\right) \qquad \text{becomes} \qquad K_m\left(1 + \frac{K_a}{[H^+]}\right) = K_m' \quad .$$

[3.35]

There are seemingly countless possible models of the pH-dependence of Michaelis-Menten and more complex enzymic reactions. The book by Roberts[7] is especially good on this topic.

Q: Draw a series of Michaelis-Menten plots for a single-substrate enzyme that has an ionizable substrate with a single pK_a of 7.0. Suppose that the forward catalytic breakdown rate constant (see Section 3.2.3) of the enzyme is 1000 s^{-1}, $[E]_0 = 10^{-6}$ M, and $K_m = 1.0$ mM.

A: A consistent enzyme reaction scheme is that given in Eqn [3.30]; and the mathematical function is given by Eqn [3.31].

```
Clear[Subscript, pH];
```

$$v_0[s\_] := \frac{V_{max}\ s}{K_m\left(1 + \frac{10^{-pH}}{K_a}\right) + s} ;$$

The following parameters are given:

$$k_{cat,f} = 1.0 \times 10^3 ;$$
$$e_0 = 1.0 \times 10^{-6} ;$$
$$K_m = 1.0 \times 10^{-3} ;$$
$$pK_a = 7.0 ;$$

From these we can determine the V_{max} and K_a values,

$$K_a = 10^{-pK_a} ;$$
$$V_{max} = k_{cat,f}\ e_0 ;$$

and a plot of Eqn [3.31] as a function of substrate concentration from pH $4 - 6$.

```
plotTable = Table[v₀[s] /. pH -> 3.5 + i * 0.5, {i, 5}];
Plot[Evaluate[plotTable], {s, 0, 100 Kₘ},
  AxesLabel → {"[s] (M)", "v₀[s] (M s⁻¹)"}];
```

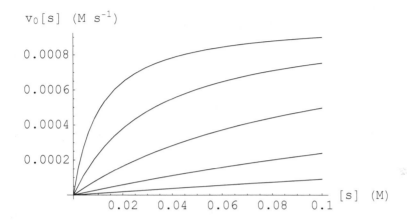

Figure 3.5. The effects of pH on the kinetics of a particular enzyme. Note that in moving from the upper to the lower curve the rates are at pH 6, 5.5, 5, 4.5, and 4.

3.8 A Simple Model of the Urea Cycle

The previous Sections of this chapter were concerned mainly with models of enzymes in isolation. Specifically, the concern was how to derive steady-state rate equations for various mechanisms and how to relate the unitary rate constants of these mechanisms to the steady-state parameters, the latter usually being the only parameters that have been measured experimentally. Thus, having described how to model individual enzymes we now turn to the simulation of metabolic pathways. As a simple example, a model of the urea cycle (Figure 3.6) is developed.

The model presented here was originally developed[8] to study the possible effects on metabolite concentrations of changing various kinetic parameters of the enzymes of the urea cycle. The concentrations of the metabolic intermediates are known to be affected in inborn errors of the enzymes of the urea cycle. The most dramatic clinical signs arise from an overall slowing of flux through the cycle and hence of a buildup of free ammonia in the body. The high ammonia concentrations lead to nausea, vomiting, loss of consciousness, convulsions, and ultimately death.

The model as it was conceived was the first to attempt to simulate the kinetics of a metabolic pathway that is subjected to an inborn error of one of the enzymes. At the time it was developed, the use of unitary rate constants led to very slow simulations of metabolic outcomes; in fact, using a Univac 1108 that had 128 K of RAM, the simulation of 10 min of a time course took 10 min of central processing unit (cpu) time. How times change! But as was shown in 1977, for most purposes the urea cycle can be simulated by using only the steady-state equations; this avoids the stiffness (see Section 1.6) that consideration of the pre-steady-state phases of the enzymic reactions imposes on the computation.

The urea cycle model consists of four enzyme reaction schemes: arginase, ornithine carbamoyl transferase, argininosuccinate lyase, and argininosuccinate synthetase. Rate equations and kinetic parameters for the first two reactions have been determined above in Sections 3.4 and 3.5. The mechanisms of the latter two enzymes are given in the exercises at the end of this chapter; and it is left as an exercise for you the reader to verify the rate equations and kinetic parameters used in the following model.

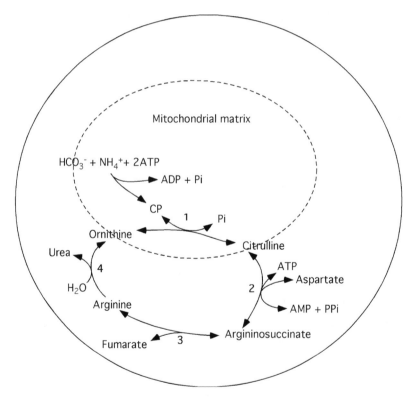

Figure 3.6. The urea cycle of the mammalian hepatocyte.

Q: Assemble the differential equations for each of the enzymes of the urea cycle and simulate a time course of the operation of the cycle for a period of 1 h.

A: The model is based on the urea cycle that operates in human liver. The steady-state output of urea is set to the rate that is found for the 'average' adult human, namely, 6.6×10^{-5} mol L^{-1} s^{-1}. Since the steady-state concentrations of the intermediates of the cycle develop rapidly, the efflux rate of other metabolites that are peripheral to the cycle must ultimately have this value as well.

To simulate this system we first define the rate equations for the four main enzymes.[8]

Ornithine carbamoyl transferase (OCT)

$e_{oct} = 2.6 \times 10^{-6};$

$k_{1,oct} = 1.7 \times 10^{7};$

$k_{-1,oct} = 63;$

$k_{2,oct} = 2.1 \times 10^{6};$

$k_{-2,oct} = 1.0 \times 10^{3};$

$k_{3,oct} = 3 \times 10^{3};$

$k_{-3,oct} = 9.0 \times 10^{4};$

$k_{4,oct} = 2.6 \times 10^{3};$

$k_{-4,oct} = 5.0 \times 10^{5};$

$v_{oct}[t\_] := e_{oct} \dfrac{1}{denom_{oct}} (cp[t] \, o[t] \, k_{1,oct} \, k_{2,oct} \, k_{3,oct} \, k_{4,oct} -$
$\quad c[t] \, p[t] \, k_{-4,oct} \, k_{-3,oct} \, k_{-2,oct} \, k_{-1,oct});$

$denom_{oct} := c[t] \, k_{-3,oct} \, k_{-2,oct} \, k_{-1,oct} + p[t] \, (k_{-4,oct} \, k_{-2,oct} \, k_{-1,oct} +$
$\quad c[t] \, (k_{-4,oct} \, k_{-3,oct} \, k_{-2,oct} + k_{-4,oct} \, k_{-3,oct} \, k_{-1,oct}) + k_{-4,oct} \, k_{-1,oct}$
$\quad\quad k_{3,oct} + o[t] \, (c[t] \, k_{-4,oct} \, k_{-3,oct} \, k_{2,oct} + k_{-4,oct} \, k_{2,oct} \, k_{3,oct})) +$
$\quad k_{-2,oct} \, k_{-1,oct} \, k_{4,oct} + k_{-1,oct} \, k_{3,oct} \, k_{4,oct} +$
$\quad o[t] \, k_{2,oct} \, k_{3,oct} \, k_{4,oct} +$
$\quad cp[t] \, (c[t] \, k_{-3,oct} \, k_{-2,oct} \, k_{1,oct} + k_{-2,oct} \, k_{1,oct} \, k_{4,oct} +$
$\quad\quad k_{1,oct} \, k_{3,oct} \, k_{4,oct} + o[t] \, (c[t] \, k_{-3,oct} \, k_{1,oct} \, k_{2,oct} +$
$\quad\quad k_{1,oct} \, k_{2,oct} \, k_{3,oct} + k_{1,oct} \, k_{2,oct} \, k_{4,oct}))$

Argininosuccinate synthetase (ASS)

$e_{ass} = 4.0 \times 10^{-6};$

$k_{1,ass} = 2.4 \, 10\text{\textasciicircum}5;$

$k_{-1,ass} = 2.3;$

$k_{2,ass} = 3.5 \, 10\text{\textasciicircum}5;$

$k_{-2,ass} = 10.0;$

$k_{3,ass} = 4.8 \, 10\text{\textasciicircum}5;$

$k_{-3,ass} = 10.0;$

$k_{4,ass} = 2.0 \, 10\text{\textasciicircum}1;$

$k_{-4,ass} = 8.9 \, 10\text{\textasciicircum}5;$

$k_{5,ass} = 5.0 \, 10\text{\textasciicircum}1;$

$k_{-5,ass} = 6.4\ 10^5;$

$k_{6,ass} = 5.0\ 10^1;$

$k_{-6,ass} = 1.7\ 10^5;$

$$v_{ass}[t\_] := e_{ass}\ \frac{1}{denom_{ass}}$$

$(k_{1,ass}\ k_{2,ass}\ k_{3,ass}\ k_{4,ass}\ k_{5,ass}\ k_{6,ass}\ c[t]\ atp[t]\ asp[t] -$
$\quad k_{-1,ass}\ k_{-2,ass}\ k_{-3,ass}\ k_{-4,ass}\ k_{-5,ass}\ k_{-6,ass}\ pp[t]\ amp[t]\ as[t]);$

$denom_{ass} := k_{-1,ass}\ k_{-2,ass}\ k_{5,ass}\ k_{6,ass}\ (k_{-3,ass} + k_{4,ass}) +$

$\quad k_{1,ass}\ k_{-2,ass}\ k_{-3,ass}\ k_{-4,ass}\ k_{6,ass}\ c[t]\ pp[t] +$

$\quad k_{1,ass}\ k_{-2,ass}\ k_{5,ass}\ k_{6,ass}\ (k_{-3,ass} + k_{4,ass})\ c[t] +$

$\quad k_{-1,ass}\ k_{3,ass}\ k_{4,ass}\ k_{5,ass}\ k_{-6,ass}\ asp[t]\ as[t] +$

$\quad k_{-1,ass}\ k_{3,ass}\ k_{4,ass}\ k_{5,ass}\ k_{6,ass}\ asp[t] +$

$\quad k_{1,ass}\ k_{2,ass}\ k_{-3,ass}\ k_{-4,ass}\ k_{6,ass}\ c[t]\ atp[t]\ pp[t] +$

$\quad k_{1,ass}\ k_{2,ass}\ k_{5,ass}\ k_{6,ass}\ (k_{-3,ass} + k_{4,ass})\ c[t]\ atp[t] +$

$\quad k_{1,ass}\ k_{-2,ass}\ k_{-3,ass}\ k_{-4,ass}\ k_{-5,ass}\ c[t]\ pp[t]\ amp[t] +$

$\quad k_{1,ass}\ k_{3,ass}\ k_{4,ass}\ k_{5,ass}\ k_{6,ass}\ c[t]\ asp[t] +$

$\quad k_{2,ass}\ k_{3,ass}\ k_{4,ass}\ k_{5,ass}\ k_{-6,ass}\ atp[t]\ asp[t]\ as[t] +$

$\quad k_{2,ass}\ k_{3,ass}\ k_{4,ass}\ k_{5,ass}\ k_{6,ass}\ atp[t]\ asp[t] +$

$\quad k_{-1,ass}\ k_{3,ass}\ k_{4,ass}\ k_{-5,ass}\ k_{-6,ass}\ asp[t]\ amp[t]\ as[t] + k_{1,ass}\ k_{2,ass}$

$\quad k_{3,ass}\ (k_{4,ass}\ k_{5,ass} + k_{4,ass}\ k_{6,ass} + k_{5,ass}\ k_{6,ass})\ c[t]\ atp[t]\ asp[t] +$

$\quad k_{1,ass}\ k_{2,ass}\ k_{3,ass}\ k_{-4,ass}\ k_{6,ass}\ c[t]\ atp[t]\ asp[t]\ pp[t] +$

$\quad k_{-1,ass}\ k_{-2,ass}\ k_{-3,ass}\ k_{-4,ass}\ k_{6,ass}\ pp[t] +$

$\quad k_{1,ass}\ k_{2,ass}\ k_{3,ass}\ k_{4,ass}\ k_{-5,ass}\ c[t]\ atp[t]\ asp[t]\ amp[t] +$

$\quad k_{-1,ass}\ k_{-2,ass}\ k_{5,ass}\ k_{-6,ass}\ (k_{-3,ass} + k_{4,ass})\ as[t] +$

$\quad k_{1,ass}\ k_{2,ass}\ k_{-3,ass}\ k_{-4,ass}\ k_{-5,ass}\ c[t]\ atp[t]\ pp[t]\ amp[t] +$

$\quad k_{-1,ass}\ k_{-2,ass}\ k_{-3,ass}\ k_{-4,ass}\ k_{-5,ass}\ pp[t]\ amp[t] +$

$\quad k_{2,ass}\ k_{3,ass}\ k_{4,ass}\ k_{-5,ass}\ k_{-6,ass}\ atp[t]\ asp[t]\ amp[t]\ as[t] +$

$\quad k_{-1,ass}\ k_{-2,ass}\ k_{-3,ass}\ k_{-4,ass}\ k_{-6,ass}\ pp[t]\ as[t] +$

$\quad k_{2,ass}\ k_{-3,ass}\ k_{-4,ass}\ k_{-5,ass}\ k_{-6,ass}\ atp[t]\ pp[t]\ amp[t]\ as[t] +$

$\quad k_{-1,ass}\ k_{-2,ass}\ k_{-5,ass}\ k_{-6,ass}\ (k_{-3,ass} + k_{4,ass})\ amp[t]\ as[t] +$

$\quad k_{-1,ass}\ k_{3,ass}\ k_{-4,ass}\ k_{-5,ass}\ k_{-6,ass}\ asp[t]\ pp[t]\ amp[t]\ as[t] +$

$\quad k_{-4,ass}\ k_{-5,ass}\ k_{-6,ass}$

$\quad (k_{-1,ass}\ k_{-2,ass} + k_{-1,ass}\ k_{-3,ass} + k_{-2,ass}\ k_{-3,ass})\ pp[t]\ amp[t]\ as[t] +$

$\quad k_{1,ass}\ k_{2,ass}\ k_{3,ass}\ k_{-4,ass}\ k_{-5,ass}\ c[t]\ atp[t]\ asp[t]\ pp[t]\ amp[t] +$

$\quad k_{2,ass}\ k_{3,ass}\ k_{-4,ass}\ k_{-5,ass}\ k_{-6,ass}\ atp[t]\ asp[t]\ pp[t]\ amp[t]\ as[t]$

Argininosuccinate lyase (ASL)

$e_{as} = 2.2 \times 10^{-6};$

$k_{1,as} = 2.7 \times 10^6;$

$k_{-1,as} = 7.0 \times 10^1;$

$k_{2,as} = 7.5 \times 10^1;$

$k_{-2,as} = 1.5 \times 10^6$;

$k_{3,as} = 1.1 \times 10^3$;

$k_{-3,as} = 7.0 \times 10^5$;

$$v_{as}[t\_] := e_{as} \frac{k_{1,as} \, k_{2,as} \, k_{3,as} \, as[t] - k_{-3,as} \, k_{-2,as} \, k_{-1,as} \, a[t] \, f[t]}{denom_{as}};$$

$denom_{as} := f[t] \, k_{-2,as} \, k_{-1,as} +$
$\quad a[t] \, (f[t] \, k_{-3,as} \, k_{-2,as} + k_{-3,as} \, k_{-1,as} + k_{-3,as} \, k_{2,as}) + k_{-1,as} \, k_{3,as} +$
$\quad k_{2,as} \, k_{3,as} + as[t] \, (f[t] \, k_{-2,as} \, k_{1,as} + k_{1,as} \, k_{2,as} + k_{1,as} \, k_{3,as});$

Arginase

$e_{arg} = 8.9 \times 10^{-6}$;

$k_{1,arg} = 1.0 \times 10^7$;

$k_{-1,arg} = 5.4 \times 10^4$;

$k_{2,arg} = 5.3 \times 10^3$;

$k_{3,arg} = 3.0 \times 10^4$;

$k_{-3,arg} = 1.0 \times 10^7$;

$$v_{arg}[t\_] := e_{arg} \frac{k_{1,arg} \, k_{2,arg} \, k_{3,arg} \, a[t]}{denom_{arg}};$$

$denom_{arg} := o[t] \, (k_{-3,arg} \, k_{-1,arg} + k_{-3,arg} \, k_{2,arg}) +$
$\quad k_{-1,arg} \, k_{3,arg} + k_{2,arg} \, k_{3,arg} + a[t] \, (k_{1,arg} \, k_{2,arg} + k_{1,arg} \, k_{3,arg});$

Next we define the rate equations for the co-substrates and products that are peripheral to the cycle proper. For these peripheral reactions we assume simple first-order kinetics, as follows.

$v_{atp}[t\_] := k_{atp} \, atppool[t];$	$k_{atp} = 6.6 \times 10^{-2};$
$v_{pp}[t\_] := k_{pp} \, pp[t];$	$k_{pp} = 6.6 \times 10^{-2};$
$v_f[t\_] := k_f \, f[t];$	$k_f = 6.6 \times 10^{-2};$
$v_{cp}[t\_] := k_{cp} \, ampool[t];$	$k_{cp} = 6.6 \times 10^{-2};$
$v_{asp}[t\_] := k_{asp} \, asppool[t];$	$k_{asp} = 6.6 \times 10^{-2};$
$v_{amp}[t\_] := k_{amp} \, amp[t];$	$k_{amp} = 6.6 \times 10^{-2};$
$v_p[t\_] := k_p \, p[t];$	$k_p = 6.6 \times 10^{-2};$

Having defined the rate equations, we are now in a position to set up the system of differential equations which make up the model.

$eqn_1 := c'[t] == v_{oct}[t] - v_{ass}[t];$

$eqn_2 := a'[t] == v_{as}[t] - v_{arg}[t];$

```
eqn₃ := u'[t] == v_arg[t];
eqn₄ := atp'[t] == v_atp[t] - v_ass[t];
eqn₅ := pp'[t] == -v_pp[t] + v_ass[t];
eqn₆ := f'[t] == -v_f[t] + v_as[t];
eqn₇ := as'[t] == v_ass[t] - v_as[t];
eqn₈ := o'[t] == v_arg[t] - v_oct[t];
eqn₉ := cp'[t] == v_cp[t] - v_oct[t];
eqn₁₀ := asp'[t] == v_asp[t] - v_ass[t];
eqn₁₁ := amp'[t] == -v_amp[t] + v_ass[t];
eqn₁₂ := p'[t] == -v_p[t] + v_oct[t];
```

In this model we assume that the pool concentrations of ATP, AMP, and aspartate are kept constant by some external processes. These concentrations can be thought of as 'external' parameters and they are assigned the following values:

```
atppool[t] = 1.0 × 10⁻⁴;
ampool[t] = 1.0 × 10⁻⁴;
asppool[t] = 1.0 × 10⁻⁴;
```

Now we solve this system of equations by using **NDSolve**

```
sol = NDSolve[{eqn₁, eqn₂, eqn₃, eqn₄, eqn₅, eqn₆, eqn₇,
    eqn₈, eqn₉, eqn₁₀, eqn₁₁, eqn₁₂, c[0.] == 1.0 × 10⁻⁷,
    as[0.] == 1.0 × 10⁻⁵, a[0.] == 1.0 × 10⁻⁷, o[0.] == 4.5 × 10⁻⁴,
    u[0.] == 1.0 × 10⁻⁵, cp[0.] == 1.0 × 10⁻⁴, atp[0.] == 1.0 × 10⁻³,
    asp[0.] == 1.0 × 10⁻³, pp[0.] == 1.0 × 10⁻⁵,
    amp[0.] == 1.0 × 10⁻⁵, f[0.] == 1.0 × 10⁻⁵, p[0.] == 1.0 × 10⁻⁵},
    {c[t], a[t], u[t], atp[t], pp[t], f[t], as[t], o[t], cp[t],
    asp[t], amp[t], p[t]}, {t, 0.0, 600}, AccuracyGoal → 10];
```

and then plot the results as follows:

```
graph1 = Plot[Evaluate[o[t] /. sol],
    {t, 0, 600}, PlotRange -> {0, 0.00045},
    AxesLabel → {"Time (s)", "[orn] (moles L^-1)"}];
```

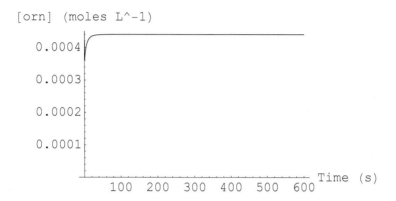

Figure 3.7. Time course of ornithine concentration in the urea cycle simulation.

```
graph2 = Plot[Evaluate[c[t] /. sol], {t, 0, 600}, PlotRange → All,
    AxesLabel → {"Time (s)", "[cit] (moles L^-1)"}];
```

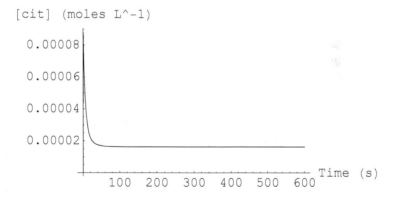

Figure 3.8. Time course of citrulline concentration in the urea cycle simulation.

```
graph3 = Plot[Evaluate[a[t] /. sol], {t, 0, 600}, PlotRange → All,
   AxesLabel → {"Time (s)", "[arg] (moles L^-1)"}];
```

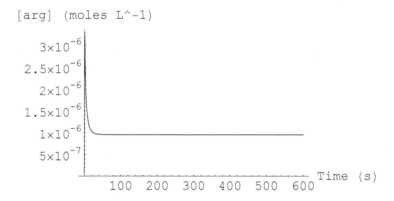

Figure 3.9. Time course of arginine concentration in the urea cycle simulation.

```
graph4 = Plot[Evaluate[u[t] /. sol], {t, 0, 600}, PlotRange → All,
   AxesLabel → {"Time (s)", "[ure] (moles L^-1)"}];
```

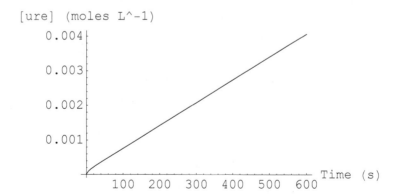

Figure 3.10. Time course of urea concentration in the urea cycle simulation.

```
Show[{graph1, graph2, graph3, graph4}];
```

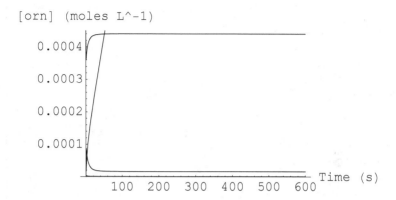

Figure 3.11. Combined plot of the previous four graphs showing ornithine (upper curve), citrulline (lower curve), and urea (the middle curve which rapidly rises out of the envelope of the graph).

3.9 Conclusions

On first view this *Mathematica* model of the urea cycle appears to be very complicated. With the detailed rate equations is there not a large scope for typographical errors and mistakes in parameter values? And, this is only a very simple metabolic model. What happens when we start to explore more complex metabolic systems?

Unfortunately the nature of metabolic systems is that they are very complicated and any model which hopes to capture the ways that these systems operate will inevitably be complex. As we saw in the urea cycle model, a large part of the modelling process involves the construction of appropriate rate equations. This part of the modelling process is by far the hardest and most complicated. Unfortunately, for most enzymes it is not possible to find a rate equation in the literature that can be simply inserted into the model that is being developed.

Most rate equations have been developed in order to understand a particular experimental situation. For example, the rate of an enzyme may have only been characterized for the reaction in one direction using only initial velocity data. In this case, the model that is fitted to the data will not include the effects of the products on the rate. While this model can be entirely appropriate for the experimental situation, it is not appropriate for a model of a 'real' cellular system. In such a system there will also be a variety of effectors that play a role in the enzyme's activity *in situ*. Hence, rate equations for modelling metabolic systems need to take into account the effects of substrates, products, and all important effectors. Developing the appropriate model requires an appreciation of the available literature on the enzyme and also its role in the wider metabolic scheme. Knowing what aspects of the reaction mechanism to include in the model, and how to incorporate data obtained from a wide range of laboratories and under a wide range of experimental conditions, forms a major part of the "art" of metabolic modelling.

Fortunately, once the appropriate reaction scheme has been determined, the mechanics of deriving rate equations and determining the relationships between steady-state parameters and unitary rate constants is largely taken care of by *Mathematica*. Thus the symbolic algebraic capabilities of *Mathematica* as well as the **RateEquation** function allows these processes to be carried out essentially free of error.

The second major task of model development is the formulation of the systems of differential equations that make up the model. Although this was a relatively simple task for the urea cycle model, it becomes a possible source of error when we deal with larger metabolic systems. In the next chapter, however, we show how entire metabolic systems can be represented in a compact form using vectors and matrices. This approach greatly simplifies the handling of metabolic systems in *Mathematica* as well as providing important new tools, such as the automatic derivation of differential equations, for modelling metabolic systems in normal and disease states.

3.10 Exercises

3.10.1

(1) Adjust the values of V_{max} and K_m in the worked example (question) in Section 3.2.1 to observe what happens to the shape and the duration of the simulated time course for both substrate and product. (2) Next, assume that this reaction produces two molecules of P from each molecule of A. What happens to the shape of the progress curves for both A and P? Hint: change the conservation of mass equations.

3.10.2

(1) For the arginase reaction (see the first question in Section 3.2.4), try substituting the values of the unitary rate constants back into the steady-state expressions to check that they do indeed return the original steady-state values. (2) Suppose that the temperature of an assay mixture is raised so that the turnover number of the enzyme is increased by a factor of 2. If all the other steady-state parameters remain unchanged, calculate a set of unitary rate constants that are consistent with these values.

3.10.3

The reaction mechanism for the urea cycle enzyme, argininosuccinate lyase, is

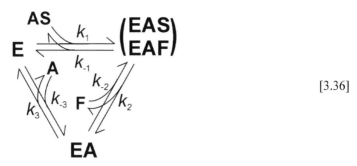

$$[3.36]$$

where AS, F, and A denote argininosuccinate, fumarate, and arginine, respectively. Derive the steady-state rate equation for this enzyme using the function **RateEquation** as in Section 3.5.

3.10.4

Use the methods described in Section 3.5 to derive the expressions for the standard steady-state kinetic parameters for arginase that has the mechanism shown in Eqn [3.17].

3.10.5

In Section 3.7 we examined the effects of pH on the kinetics of an enzyme. For the worked example in this section, (1) alter the range of pH values for which the Michaelis-Menten equation is plotted. Notice the extent to which V_{max} is likely to be able to be accurately inferred from the graphs; and (2) what happens to the apparent K_m as the pH is varied?

3.10.6

Verify that the unitary rate constants used for argininosuccinate lyase that were used in the model of the urea cycle presented in Section 3.8 are consistent with the following steady-state parameters and the overall equilibrium constant for the reaction: $k_{cat}^f = 70$ mol L^{-1} s^{-1}; $K_{m, AS} = 5 \times 10^{-5}$ mol L^{-1}; $K_{m, F} = 1 \times 10^{-4}$ mol L^{-1}; $K_{m, A} = 1 \times 10^{-4}$ mol L^{-1}; $K_{eq} = 3.2 \times 10^{-3}$ mol L^{-1}. The reaction mechanism for argininosuccinate lyase is given in Exercise 3.3.

3.10.7

The reaction mechanism for argininosuccinate synthetase is

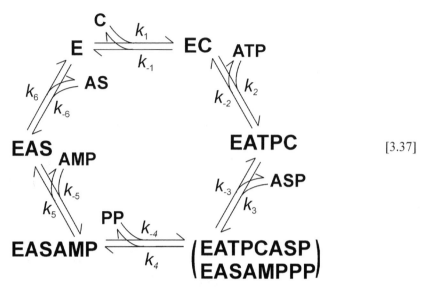

[3.37]

where C, ATP, ASP, PP, AMP, and AS denote citrulline, ATP, aspartate, pyrophosphate, AMP, and argininosuccinate, respectively. Verify that the unitary rate constants used in the model of the urea cycle model presented in Section 3.8 are consistent with the following steady-state kinetic parameters and the overall equilibrium constant for the reaction: $k_{cat}^f = 11.2 \text{ mol}^2 \text{ L}^{-2} \text{ s}^{-1}$; $K_{m,C} = 4.6 \times 10^{-5} \text{ mol L}^{-1}$; $K_{m,ATP} = 3.2 \times 10^{-4} \text{ mol L}^{-1}$; $K_{m,ASP} = 3.5 \times 10^{-5} \text{ mol L}^{-1}$; $K_{eq} = 8.9$; $K_{m,CAS} = 3.0 \times 10^{-4} \text{ mol L}^{-1}$; $K_{m,CPP} = 1.0 \times 10^{-4} \text{ mol L}^{-1}$; $K_{i,ATPAMP} = 3.5 \times 10^{-4} \text{ mol L}^{-1}$.

3.10.8

Clinically, arginase deficiency leads to profound hyperammonemia. It has been found that in the patients who survive early infancy, the maximal activity of the enzyme is around 3% of the normal value. Assuming that this implies that there is only 3% of the normal enzyme concentration, simulate the operation of the human urea cycle under this pathological condition. Comment on the concentrations of the intermediates and whether they attain steady states. In the event of the latter not occurring, speculate on the likely outcome for the particular metabolite(s) and hence the clinical status of the patient.

3.10.9

There are basically two major clinical variants or types of inborn errors of argininosuccinate lyase deficiency: (1) one condition in which there are early signs of the severe onset of hyperammonemia that leads to death in infancy; and (2) a milder course of disease with survival into adolescence. The former appears to result from a low concentration of normal enzyme, while the latter patients probably have an enzyme with lower substrate affinity.

(1) Suppose that in the first case $[E]_0$ is 3% of its normal value. Simulate the operation

of the urea cycle for a period of 1 h. Comment on dramatic features of the simulated time course.

(2) Simulate the operation of the urea cycle in which the affinity of the enzyme for argininosuccinate lyase is reduced to 3% of its normal value. Again, comment on any dramatic findings.   Does the system attain a new steady state of metabolite concentrations?

3.10.10

This exercise will be able to be completed only after reading Chapters 4 and 5. Apply MCA (Chapter 5) to the model of the urea cycle described in Section 3.8 and determine the values of the flux control coefficients for each enzyme. Comment on the occurrence of metabolic resistance in these simulations; this occurs in situations where the affinity of the enzyme for one of its substrates is diminished and the steady-state concentration of the substrate(s) rises to a new value thus overcoming the blockage of the pathway.

3.11 References

1. Conigrave, A.C. and Morris, M.B. (1998) A 96-well plate assay for the study of calmodulin-activated Ca2+-pumping ATPase from red-cell membranes. *Biochem. Ed.* **26**, 176-181.
2. Mulquiney, P.J. and Kuchel, P.W. (1997) Model of the pH-dependence of the concentration of complexes involving metabolites, haemoglobin and magnesium ions in the human erythrocyte. *Eur. J. Biochem.* **245**, 71-83.
3. Mulquiney, P.J. and Kuchel, P.W. (1997) Free magnesium-ion concentration in erythrocytes by [31] P NMR: the effect of metabolite haemoglobin interactions. *NMR Biomed.* **245**, 71-83.
4. Cornish-Bowden, A. (1995) *Fundamentals of Enzyme Kinetics*. Portland Press, London.
5. Kuchel, P.W. and Ralston, G.B. (Eds.) (1998) *Schaum's Outline of Theory and Problems of Biochemistry*. 2nd ed. McGraw-Hill, New York, pp. 251-289.
6. Plowman, K.M. (1972) *Enzyme Kinetics*. McGraw-Hill, New York.
7. Roberts, D.V. (1977) *Enzyme Kinetics*, Cambridge University Press, New York.
8. Kuchel, P.W., Roberts, D.V., and Nichol, L.W. (1977) The simulation of the urea cycle: correlation of effects due to inborn errors in the catalytic properties of the enzymes with clinical-biochemical observations. *Aust. J. Exp. Biol. Med. Sci.* **55**, 309-326.

4 Advanced Simulation of Metabolic Pathways

4.1 Introduction

In the previous chapter we examined the kinetic behaviour of reactions catalyzed by individual isolated enzymes and built up to systems of rate equations that describe metabolic pathways. In this Chapter and the next we show how entire metabolic systems can be represented in compact form using vectors and matrices. Although for those not familiar with matrices this process may at first seem obscure and unnecessarily difficult, the payoff is greater ease of model construction and analysis.

4.2 Simulating the Time-Dependent Behaviour of Multi-enzyme Systems

Before we begin our matrix approach to modelling, it is worthwhile re-examining the approach we have used so far. This is best illustrated with the following simple example.

Q: Determine the time dependence of the concentrations of reactants in the two-reaction sequence,

$$S_1 \underset{}{\overset{v_1}{\rightleftharpoons}} S_2 \underset{}{\overset{v_2}{\rightleftharpoons}} S_3 \tag{4.1}$$

where v_1 and v_2 are described by simple reversible Michaelis-Menten rate equations (Section 2.3.3) with all V_{max} values being 1 mmol L^{-1} h^{-1} and all K_m values being 1 mmol L^{-1}. Suppose that the initial concentrations are $S_1[0] = 1$ mmol L^{-1}, $S_2[0] = 0$, and $S_3[0] = 0$.

A: The analysis begins by defining the rate expressions and assigning values to the various parameters. Then the differential equations are solved numerically with **NDSolve** yielding a result that is stored as an **InterpolatingFunction** (Section 1.4.6).

$$\mathbf{v}[1] := \frac{\dfrac{V_{max,1,f} \, s_1[t]}{K_{m,1,s_1}} - \dfrac{V_{max,1,r} \, s_2[t]}{K_{m,1,s_2}}}{1 + \dfrac{s_1[t]}{K_{m,1,s_1}} + \dfrac{s_2[t]}{K_{m,1,s_2}}} ;$$

$$\mathbf{v}[2] := \frac{\dfrac{V_{max,2,f} \, s_2[t]}{K_{m,2,s_2}} - \dfrac{V_{max,2,r} \, s_3[t]}{K_{m,2,s_3}}}{1 + \dfrac{s_2[t]}{K_{m,2,s_2}} + \dfrac{s_3[t]}{K_{m,2,s_3}}} ;$$

```
Vmax,1,f = 1; Vmax,1,r = 1;
Vmax,2,f = 1; Vmax,2,r = 1;
Km,1,s1 = 1; Km,1,s2 = 1;
Km,2,s2 = 1; Km,2,s3 = 1;

timecourse =
  NDSolve[{
    s1'[t] == -v[1],
    s2'[t] == v[1] - v[2],
    s3'[t] == v[2],
    s1[0] == 1, s2[0] == 0, s3[0] == 0},
    {s1, s2, s3}, {t, 0, 10}]
{{s1 → InterpolatingFunction[{{0., 10.}}, <>],
  s2 → InterpolatingFunction[{{0., 10.}}, <>],
  s3 → InterpolatingFunction[{{0., 10.}}, <>]}}
```

Hence, plotting the time course of all three metabolites over the entire 10 h of simulated time we obtain the following:

```
Plot[Evaluate[{s1[t], s2[t], s3[t]} /. timecourse], {t, 0, 10},
  AxesLabel -> {"Time (h)", "Concentration (mM)"},
  PlotRange -> {0, 1}];
```

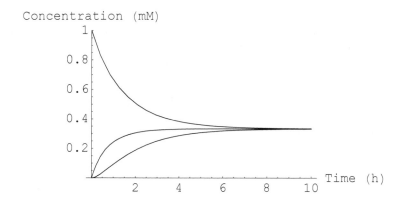

Figure 4.1. Time course of the reaction scheme shown in Eqn [4.1]. Upper curve is S_1, middle curve S_2, and lower curve S_3.

Recall that we can evaluate the concentration of the three metabolites with the following command:

```
{s₁[t], s₂[t], s₃[t]} /. timecourse /. t -> 1

{{0.665123, 0.242544, 0.0923338}}
```

4.3 Using Matrix Notation in Simulating Metabolic Pathways

For large metabolic networks, it is often simpler to express the set of simultaneous differential equations that describe the reaction system in matrix form. The use of matrix notation is also useful for analysing the existence and stability of steady states and for performing MCA (see Chapter 5).

Thus the set of differential equations describing the rate of production and utilization of all metabolites, S_i, in a metabolic network of reactions is given by

$$\frac{dS_i}{dt} = \sum_{j=1}^{r} n_{ij} v_j \; , \qquad [4.2]$$

where v_j (j = 1,...,r) is the rate of reaction j, and n_{ij} is the stoichiometric coefficient with i and j referring to the metabolite and reaction, respectively.

In matrix notation, Eqn [4.2] is given by

$$\frac{dS}{dt} = N v \; , \qquad [4.3]$$

where **v** and **S** denote the vectors of reaction rate expressions and concentrations, respectively, while **N** is termed the stoichiometry matrix with each element being n_{ij}. The stoichiometry matrix has one row for each reactant and one column for each reaction in the scheme. The number at the intersection of a row and a column defines the number of molecules of that reactant to engage in the specified elementary reaction, namely, its stoichiometry. Although this representation may seem overly complex at present, expressions like Eqn [4.2] and [4.3] are almost invariably much simpler than they look at first sight because in a large model most of the n_{ij} are zero and nearly all the rest are either – 1 or 1.

To analyse systems of differential equations that are represented in matrix notation requires a moderate amount of computer programming. However, once the general procedure has been set up it can be used for any reaction scheme. Because of its general usefulness we have written an add-on package called **MetabolicControl-Analysis** that contains the matrix representation of reaction schemes and several other useful functions that are described below and in Chapter 5. The details of the programming are not given in the text but the programs in Appendix 2 contain many comment statements to help you, the reader, understand the algorithms. Before this package is used for the first time, the reader must evaluate all the Cells in Appendix 2 so that the appropriate .m file is created and the functions can be called in a *Mathematica* session.

<<	read in the add –
MetabolicControlAnalysis	on package MetabolicControlAnalysis
NDSolveMatrix[*S*, *N*, *v*,	Uses the function **NDSolve** to find a numerical
initial conditions, {*t*, *tmin*, *tmax*}]	solution for the metabolite concentrations,
	S, with time in the range *tmin* to *tmax*,
	for a system of ordinary differential
	equations defined by the matrices *S*, *N*,
	and *v*, and subject to the *initial conditions*

Reading in the package MetabolicControlAnalysis.

Q: By using the matrix representation of a reaction scheme, plot the 10 h time course of the concentrations of the reactants in the following linear sequence,

$$S_1 \xrightarrow{v_1} S_2 \xrightarrow{v_2} S_3 \xrightarrow{v_3} S_4 \qquad\qquad [4.4]$$

where the v_i are simple irreversible Michaelis-Menten rate equations (Section 2.3.1) with all V_{max} values set to 1 mmol L^{-1} h^{-1}; all K_m values are 1 mmol L^{-1} and $S_1[0] = 1$ mmol L^{-1}, $S_2[0] = S_3[0] = S_4[0] = 0$ mmol L^{-1}.

A: The first step in answering the question is to load the **MetabolicControl-Analysis** package.

```
<< MetabolicControlAnalysis`
```

Then the matrices and vectors of Eqn [4.3] must be defined. A simple way to define the concentration vector is as follows:

```
S̄ := Table[sᵢ[t], {i, 4}];
```

Similarly, the stoichiometry and rate equation vectors are defined as follows:

$$\tilde{N} = \begin{pmatrix} -1 & 0 & 0 \\ 1 & -1 & 0 \\ 0 & 1 & -1 \\ 0 & 0 & 1 \end{pmatrix};$$

$$\text{Do}\left[v[i] := \frac{V_{max,i,f}\, s_i[t]}{K_{m,i,s_i}\left(1 + \frac{s_i[t]}{K_{m,i,s_i}}\right)},\ \{i,\ 3\}\right];$$

$\tilde{v} := \text{Table}[v[i],\ \{i,\ 3\}];$

$V_{max,1,f} = 1;\ V_{max,2,f} = 1;\ V_{max,3,f} = 1;\ K_{m,1,s_1} = 1;\ K_{m,2,s_2} = 1;\ K_{m,3,s_3} = 1;$

With the **S**, **N**, and **v** matrices defined and the parameter values set, **NDSolveMatrix** is used to obtain the numerical solution.

```
sol = NDSolveMatrix[S̃, Ñ, ṽ,
   {s₁[0] == 1, s₂[0] == 0, s₃[0] == 0, s₄[0] == 0}, {t, 0, 10}]

{{s₁ → InterpolatingFunction[{{0., 10.}}, <>],
  s₂ → InterpolatingFunction[{{0., 10.}}, <>],
  s₃ → InterpolatingFunction[{{0., 10.}}, <>],
  s₄ → InterpolatingFunction[{{0., 10.}}, <>]}}
```

The following is a plot of all the concentrations of all the S_i that are listed in the vector **S**:

```
Plot[Evaluate[S̃ /. sol], {t, 0, 10},
   AxesLabel -> {"Time (h)", "Concentration (mM)"},
   PlotRange -> {0, 1}];
```

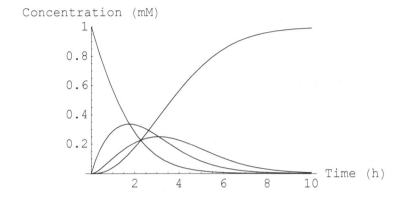

Figure 4.2. Time course of the reaction scheme shown in Eqn [4.4]. The figure shows S_1 declining from an initial concentration of 1 mM and S_4 accumulating to ~1 mM by the end of the time course. The metabolites S_2 and S_3 are seen to build up and then decline during the time course.

And here is the value of each of the reactants listed in **S** at 2 h:

```
s̄ /. t -> 2 /. sol
```

```
{{0.278465, 0.332676, 0.210329, 0.17853}}
```

The list of four numbers gives the concentrations in mM for the reactants in the order in which they are defined in **S**. Thus, S_1 is 0.28 mM, S_2 is 0.33 mM, etc.

In the worked example above the parameter values were individually specified to be 1, with units that depend on the nature of the parameter. By setting the parameter values with the **=** command, whenever *Mathematica* sees the parameter it automatically replaces it with the value. For some analyses, these parameter value assignments will need to be unset or cleared and we shall see examples of this later in Chapter 5. This can be tedious when simulating complex networks of reactions. To obviate this problem, the function **NDSolveMatrix** can accept parameter values as a separate parameter vector.

NDSolveMatrix[S, N, v, *initial conditions, {t, tmin, tmax}, p]*	Uses the function **NDSolve** to find a numerical solution for the metabolite concentrations, S, with time in the range *tmin* to *tmax*, for a system of ordinary differential equations defined by the matrices S, N, v, and the parameter matrix p

Finding numerical solutions to a system of differential equations with parameter values specified in a parameter matrix.

Q: Set all the K_m values in the previous worked example to 0.5 mmol L^{-1} and re-simulate the time course using a matrix representation of the K_m and other parameter values.

A: First the previous values should be cleared and then the matrix denoted $\bar{p}$ is specified.

```
Clear[Subscript];
```

```
p̄ = {{Vmax,1,f, 1}, {Vmax,2,f, 1}, {Vmax,3,f, 1}, {Km,1,s₁, 0.5},
   {Km,2,s₂, 0.5}, {Km,3,s₃, 0.5}} ; p̄ // MatrixForm
```

$$\begin{pmatrix} V_{max,1,f} & 1 \\ V_{max,2,f} & 1 \\ V_{max,3,f} & 1 \\ K_{m,1,s_1} & 0.5 \\ K_{m,2,s_2} & 0.5 \\ K_{m,3,s_3} & 0.5 \end{pmatrix}$$

The matrix has the parameter name in the first column and its numerical value in the second. Often, a more convenient way of inputting the parameter matrix is to do as follows:

```
p = {Vmax,1,f, Vmax,2,f, Vmax,3,f, Km,1,s₁, Km,2,s₂, Km,3,s₃};
pv = {1, 1, 1, 0.5, 0.5, 0.5};
p̄ = Transpose[{p, pv}]
```

```
{{Vmax,1,f, 1}, {Vmax,2,f, 1}, {Vmax,3,f, 1},
  {Km,1,s₁, 0.5}, {Km,2,s₂, 0.5}, {Km,3,s₃, 0.5}}
```

The numerical solution and graphical output of the system are then obtained by using

```
sol = NDSolveMatrix[S̄, N̄, v̄,
   {s₁[0] == 1, s₂[0] == 0, s₃[0] == 0, s₄[0] == 0}, {t, 0, 10}, p̄];
```

```
Plot[Evaluate[S̄ /. sol], {t, 0, 10},
   AxesLabel -> {"Time (h)", "Concentration (mM)"},
   PlotRange -> {0, 1}];
```

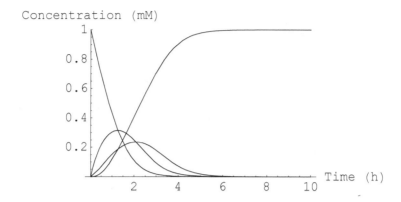

Figure 4.3. Time course of the reaction scheme shown in Eqn [4.1]. S_1 declines from an initial concentration of 1 mmol L^{-1} and S_4 accumulates to ~1 mmol L^{-1} by the end of the time course. The metabolites S_2 and S_3 are seen to build up and then decline during the time course.

Note the decline in $[S_1]$ with time, the rise to maximum concentrations and then the decline of the two intermediate species, S_2 and S_3, and the sigmoidal monotonically strictly increasing concentration of S_4.

4.4 Generating the Stoichiometry Matrix

In the above examples the task of specifying the stoichiometric matrix, **N**, was a simple one. However, for larger and more complex metabolic networks, the generation of these matrices can be tedious and error prone. The functions **StoichiometryMatrix** and **NMatrix** automate this process, thus generating the stoichiometic matrix from the list of reactions that constitute the metabolic scheme of interest.

NMatrix[*eqn, extpars*]	generates a stoichiometry matrix for the reaction system defined in the equation list *eqn;* it takes into account the fact that the parameters in the list, *extpars*, are any external parameters	
StoichiometryMatrix[*eqn, extpars*]	same as **NMatrix** except that it returns a stiochiometry matrix which has rows and columns labelled with metabolite names and reaction names, respectively	
SMatrix[*eqn, extpars*]	generates the corresponding substrate list, *S*, for the reaction system defined by *eqn* and *extpars*	
VMatrix[*eqn*]	generates the corresponding reaction velocity list, *v*, for the reaction system defined by *eqn*	

Generating stoichiometry matrices and substrate lists.

Q: Generate a stoichiometry matrix, substrate list, and reaction velocity list for the following outline model of anaerobic energy metabolism in muscle:

$$\text{Glc} + 2\,\text{ADP} + 2\,\text{Phos} \xrightarrow{v_{glyc}} 2\,\text{Lac} + 2\,\text{ATP}, \qquad [4.5]$$

$$\text{ATP} \xrightarrow{v_{atpase}} \text{ADP} + 2\,\text{Phos}, \qquad [4.6]$$

$$\text{ADP} + \text{PCr} \xrightarrow{v_{ck}} \text{ATP} + \text{Cr}, \qquad [4.7]$$

$$\text{ADP} + \text{ADP} \xrightarrow{v_{mk}} \text{ATP} + \text{AMP}, \qquad [4.8]$$

where Eqns [4.5] – [4.8] describe glycolysis, various ATPases, creatine kinase, and myokinase, respectively. Assume that the concentrations of glucose and lactate remain constant. In other words, assume that the rate of input of glucose and the rate of removal of lactate from the system are such that the concentrations of these metabolites remain constant.

A: To generate the required matrices and lists we first need to input a list that contains the details of the reaction scheme in a format that the **MetabolicControlAnalysis** package commands can recognise. This can be done with the following list of labelled **eqns**:

```
eqns = {
   {gly,       glc + 2 adp + 2 phos → 2 lac + 2 atp},
   {atpase,    atp                   →   adp +  phos},
   {ck,        adp +  pcr            ↔   atp +  cr},
   {mk,       2 adp                  ↔   atp +  amp}};
```

There are a number of features to note about the format of **eqns**. The first is that each element of the list contains the name of the reaction and then the reaction written in conventional biochemical notation. The name that is entered for each reaction is used to index the reaction rate equation. For example, by naming the first reaction **gly** we have told *Mathematica* that its associated rate equation will be called **v[gly]**. So when we define the rate equation for this reaction we will need to give it the name **v[gly]**. It is, however, not necessary to give a name to each reaction. For example, if we had failed to name the reactions, the first reaction would then be called reaction number 1 with associated rate equation **v[1]**, and so on. The other important feature in the format of **eqns** is the use of the arrow to separate the substrates from the products in each reaction. You can use any type of arrow you like to do this separation. The type that is used does not effect the final form of the stoichiometric matrix. In this example we have used one-way arrows to denote irreversible reactions and two-way arrows to denote reversible reactions, but this is only for our own benefit and it makes no difference to the generation of the stoichiometry matrix.

Having defined the equation list we are now in a position to use the functions **SMatrix**, **NMatrix**, and **VMatrix** from the **MetabolicControlAnalysis** package, to directly generate **S**, **N**, and **v**. This can be done with the following input:

S̄ = SMatrix[eqns, {glc, lac}]; S̄ // MatrixForm

$$\begin{pmatrix} \text{adp} \\ \text{amp} \\ \text{atp} \\ \text{cr} \\ \text{pcr} \\ \text{phos} \end{pmatrix}$$

In generating the list of substrates we have given a list of the 'external' metabolites as the second argument of the **SMatrix** function. These are the metabolites whose concentrations are kept constant by some external processes. Having done this, the function then picked out all six of the 'internal' metabolites and has listed them in alphabetical order in the vector. This is the order used for the rows in the following stoichiometry matrix:

Ñ = NMatrix[eqns, {glc, lac}] ; Ñ // MatrixForm

$$\begin{pmatrix} -2 & 1 & -1 & -2 \\ 0 & 0 & 0 & 1 \\ 2 & -1 & 1 & 1 \\ 0 & 0 & 1 & 0 \\ 0 & 0 & -1 & 0 \\ -2 & 1 & 0 & 0 \end{pmatrix}$$

Similarly, the columns of **N** correspond to the following vector of reaction rate expressions:

v̄ = VMatrix[eqns]; v̄ // MatrixForm

$$\begin{pmatrix} \texttt{v[gly]} \\ \texttt{v[atpase]} \\ \texttt{v[ck]} \\ \texttt{v[mk]} \end{pmatrix}$$

Note that as mentioned above, each reaction rate in **v** is indexed by the names in the first column of **eqns**.

In the absence of generating **S** it would not be immediately obvious which row corresponds to which metabolite, and which column to which reaction. Thus it is useful to be able to generate an annotated form as is shown next. The order of the reactions is simply that used in the initial **eqns** list.

```
StoichiometryMatrix[eqns, {Glc, Lac}] // MatrixForm
```

$$\begin{pmatrix}
\texttt{s}\iota\texttt{r} & \texttt{gly} & \texttt{atpase} & \texttt{ck} & \texttt{mk} \\
\texttt{adp} & -2 & 1 & -1 & -2 \\
\texttt{amp} & 0 & 0 & 0 & 1 \\
\texttt{atp} & 2 & -1 & 1 & 1 \\
\texttt{cr} & 0 & 0 & 1 & 0 \\
\texttt{glc} & -1 & 0 & 0 & 0 \\
\texttt{lac} & 2 & 0 & 0 & 0 \\
\texttt{pcr} & 0 & 0 & -1 & 0 \\
\texttt{phos} & -2 & 1 & 0 & 0
\end{pmatrix}$$

Thus it is clear, for example, that in each elementary ATPase reaction, 1 ADP molecule and 1 phos(phate) are produced, and 1 ATP is consumed.

4.5 Determining Steady-State Concentrations

In general, a metabolic pathway can be thought of as a network of biochemical reactions whose function is to take in substrates from the environment and transform them into something that is required by the organism. Thus the organism may take in fuel molecules which are used to produce ATP which is, in turn, used as the main energy 'currency' of the cell. Or alternatively, the organism may take in substrates which it uses as the building blocks for different subcellular components. Thus we can often think of a metabolic pathway as having a source substrate (such as the fuel and building block molecules) and an end product or sink (such as ATP or the subcellular components).

If the sources or sinks of a metabolic pathway do not change significantly in concentrations, or if their concentrations do not significantly effect the rates of the reaction in the metabolic pathway, then the pathway can often develop a steady state. This occurs when the concentrations of all the metabolites between the source(s) and the sink(s) do not change over time. Thus the idea of a steady state in a metabolic pathway is similar to the idea of the steady state in enzyme kinetics, a concept that was introduced in Chapter 2. Mathematically, the steady state of a metabolic pathway is described by setting the right-hand side of Eqn [4.3] to 0, i.e.,

$$\mathbf{N}\,\mathbf{v} = \mathbf{0} \; . \tag{4.9}$$

The idea of a steady state is really a mathematical abstraction that never completely occurs in reality. One reason for this is that there is never a situation when the concentrations of the sources or sinks are completely constant or have absolutely no effect on the reaction rates. A second reason is that in reality, it would take an infinite amount of time to reach a true steady state. This is because as the system moves closer to a steady state, the rate at which the system approaches the steady state becomes slower. Notwithstanding these limitations, the idea of steady state is still a very useful one. For example, routine measurements of the glycolytic metabolites in red blood cells and other cell types show that as long as the cells are incubated under the same conditions, the assayed concentrations of the metabolites remain remarkably constant over time.

To determine the steady state of a metabolic system, it is necessary to solve the set of nonlinear algebraic equations that are defined by Eqn [4.9]. This process can be tedious to program de novo, so we have provided the following function in the **MetabolicControlAnalysis** package, which carries out the operation.

SteadyState[*S*, *N*, *v*, *p*]	Uses the *Mathematica* function **Solve** to determine the solution to Eqn [4.9]. Note that like **NDSolveMatrix**, the inclusion of the parameter table *p* is optional

Solving metabolic steady states.

Q: Consider the linear three-reaction sequence described in the worked example in Section 4.3, but now assume that the concentrations of S_1 and S_4 are fixed at 1 mM by external processes. What are the steady-state concentrations of S_2 and S_3 ?

A: The first step in answering this question is to define an equation list of the reaction sequence that is given in Section 4.3.

```
eqns = { {s₁[t]  →  s₂[t]},
         {s₂[t]  →  s₃[t]},
         {s₃[t]  →  s₄[t]}};
```

Having done this, we are now in a position to define the appropriate matrices and vectors of the system. Note that when doing this we define S_1 and S_2 to be external parameters. So the relevant substrate list is

```
S̄ := SMatrix[eqns, {s₁[t] , s₄[t]}] ;
S̄ // MatrixForm
```

$$\begin{pmatrix} s_2[t] \\ s_3[t] \end{pmatrix}$$

and the relevant stoichiometry matrix and reaction list are

```
Ñ = NMatrix[eqns, {s₁[t] , s₄[t]}] ;
Ñ // MatrixForm
```

$$\begin{pmatrix} 1 & -1 & 0 \\ 0 & 1 & -1 \end{pmatrix}$$

$$\mathbf{Do}\left[\mathbf{v}[\mathbf{i}] = \frac{V_{max,i,f}\, s_i[t]}{K_{m,i,s_i}\left(1+\frac{s_i[t]}{K_{m,i,s_i}}\right)},\ \{\mathbf{i},\ 3\}\right];$$

```
v̄ := VMatrix[eqns] ;
v̄ // MatrixForm
```

$$\begin{pmatrix} \dfrac{V_{max,1,f}\, s_1[t]}{K_{m,1,s_1}\left(1+\frac{s_1[t]}{K_{m,1,s_1}}\right)} \\[2ex] \dfrac{V_{max,2,f}\, s_2[t]}{K_{m,2,s_2}\left(1+\frac{s_2[t]}{K_{m,2,s_2}}\right)} \\[2ex] \dfrac{V_{max,3,f}\, s_3[t]}{K_{m,3,s_3}\left(1+\frac{s_3[t]}{K_{m,3,s_3}}\right)} \end{pmatrix}$$

So if we specify the parameters, we can then use **SteadyState** to determine the steady state.

```
p = {s₁[t], s₄[t], Vmax,1,f, Vmax,2,f, Vmax,3,f, Km,1,s₁, Km,2,s₂, Km,3,s₃};
pv = {1, 1, 1, 1, 1, 1, 1, 1};
SteadyState[S̄, Ñ, v̄, Transpose[{p, pv}]]
```

$$\{\{s_2[t] \to 1,\ s_3[t] \to 1\}\}$$

Thus the function **SteadyState** returns a replacement rule which can be used in the usual manner to obtain the concentration of each metabolite.

Alternatively, **NDSolveMatrix** can be used to follow the time course of $[S_2]$ and $[S_3]$ from an initial concentration of 0 mM to their respective steady-state values. The graphical output shows that the steady state is attained after ~30 h.

```
sol = NDSolveMatrix[S̄, Ñ, v̄,
    {s₂[0] == 0, s₃[0] == 0}, {t, 0, 30}, Transpose[{p, pv}]];

Plot[Evaluate[S̄ /. sol], {t, 0, 30},
    AxesLabel -> {"Time (h)", "Concentration (mM)"}];
```

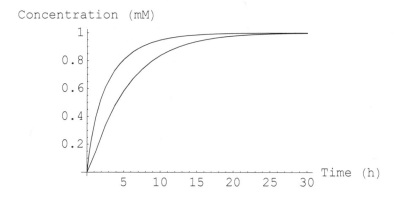

Figure 4.4. Time course of the reaction scheme shown in Eqn [4.1] with metabolites S_1 and S_4 at constant concentrations. Upper curve is S_2 and lower curve S_3.

By leaving out the V_{max} and K_m from the parameter table it is possible to obtain expressions for the steady-state concentrations of S_2 and S_3 written in terms of V_{max} and K_m. This enables visual inspection of the expressions which may help obtain an idea of which parameters most influence the steady-state concentrations. This analysis can be a prelude to the quantitative MCA that is discussed in Chapter 5.

We can implement this analysis by not including the V_{max} and K_m parameters in the parameter table that was defined at the beginning of this example. Thus,

```
p2 = {s₁[t], s₄[t]};
pv2 = {1, 1};
```

```
SteadyState[S̄, Ñ, v̄, Transpose[{p2, pv2}]]
```

$$\left\{\left\{ s_2[t] \to -\frac{K_{m,2,s_2}\, V_{max,1,f}}{V_{max,1,f} - V_{max,2,f} - K_{m,1,s_1}\, V_{max,2,f}} \,,\right.\right.$$
$$\left.\left. s_3[t] \to -\frac{K_{m,3,s_3}\, V_{max,1,f}}{V_{max,1,f} - V_{max,3,f} - K_{m,1,s_1}\, V_{max,3,f}} \right\}\right\}$$

By omitting **p** altogether, the steady state can be expressed in terms of V_{max}, K_m, S_1, and S_4.

```
SteadyState[S̄, Ñ, v̄]
```

$$\Big\{\Big\{s_2[t] \to -\frac{K_{m,2,s_2} V_{max,1,f} s_1[t]}{-K_{m,1,s_1} V_{max,2,f} + V_{max,1,f} s_1[t] - V_{max,2,f} s_1[t]},$$

$$s_3[t] \to -\frac{K_{m,3,s_3} V_{max,1,f} s_1[t]}{-K_{m,1,s_1} V_{max,3,f} + V_{max,1,f} s_1[t] - V_{max,3,f} s_1[t]}\Big\}\Big\}$$

The **SteadyState** function will often be unable to locate a solution for metabolic models that are more complicated than the one described in the above question/answer. In large metabolic models there may be no analytical solutions or the solution may be so complicated that the algorithm simply cannot determine it. In the latter cases, the function **NSteadyState** can be used to find an approximate numerical solution. This algorithm is based on Mathematica's **FindRoot** function.

NSteadyState[*S, N, v, p, init*]	Uses the *Mathematica* function **FindRoot** to determine an approximate numerical solution to Eqn [4.9] where *init* contains initial estimates of the steady state concentations in the form of a replacement rule. Note that the inclusion of the parameter table *p* is optional.

Solving metabolic steady states numerically.

Q: Calculate the steady-state concentrations of S_2 and S_3 for the metabolic scheme described in the previous question, using **NSteadyState**.

A: To use **NSteadyState** we must first set up a replacement rule containing initial estimates of the steady-state concentrations of S_2 and S_3. We can use the simulation values at 20 h that were obtained in the previous question/answer to do this.

```
ssvalues = s̄ /. t -> 30 /. sol[[1]];
rrule = Table[s̄[[i]] -> ssvalues[[i]], {i, Length[s̄]}]

{s₂[t] → 0.999664, s₃[t] → 0.997363}
```

Note that in the above input we have relied on the **Part** or **[[...]]** function.

With the initial estimate, we can refine the answer by using the function, **NSteadyState**.

```
NSteadyState[s̄, Ñ, v̄, Transpose[{p, pv}], rrule]

{s₂[t] → 1., s₃[t] → 0.999997}
```

In conclusion, we see that the numerical estimate is the same as that returned in the previous example.

$list_{[i]}$	returns the *i*th element of a list

Selecting elements of lists.

4.6 Conservation Relations

A defining feature of any metabolic system is its metabolite-conservation relationships. For example, in the human erythrocyte under normal conditions the sums of the concentrations of the adenine nucleotides ([ATP]+[ADP]+[AMP]) and the nicotinamide adenine dinucleotides ([NAD]+[NADH]) are both constant.

The stoichiometry matrix, **N**, which contains the stoichiometric structure of the model of the metabolic system, implicitly contains these conservation relationships in it. A function in the **MetabolicControlAnalysis** package called **Conservation-Relations** has been written to extract these relationships from an **N** matrix. No knowledge of the kinetic behaviour of the individual reactions is needed for the analysis so it is done by using only **S** and **N**, as follows.

ConservationRelations[S, N]	Determines nonnegative conservation relations between the metabolites S_i in the metabolic network defined by N.

Determining conservation relations between metabolites.

Q: Calculate the conservation relationships if any exist in the reaction scheme given in the question/answer in Section 4.3. Recall that the scheme can be represented by the metabolite and stoichiometry matrices.

A: This is achieved as follows:

```
S̄ = Table[sᵢ[t], {i, 4}];
```

$$\bar{N} = \begin{pmatrix} -1 & 0 & 0 \\ 1 & -1 & 0 \\ 0 & 1 & -1 \\ 0 & 0 & 1 \end{pmatrix};$$

The next function identifies any conservation relationships.

```
ConservationRelations[S̄, Ñ]
```

```
{s₁[t] + s₂[t] + s₃[t] + s₄[t]} == {Const[1]}
```

Thus, the conservation condition is such that the sum of all metabolite concentrations in the metabolic system is constant. In general, the conservation relations can also be returned in terms of a matrix, **G**.

$$\mathbf{G.S} = \mathbf{Const} ,\qquad\qquad [4.10]$$

where **Const** is a matrix of constants. The function **ConservationRelations** will return **G** if the option **GMatrix → True** is included. Hence,

```
Ḡ = ConservationRelations[S̄, Ñ, GMatrix → True];
Ḡ // MatrixForm
Ḡ.S̄ == Array[Const, Length[Ḡ]]
```

```
( 1  1  1  1 )
```

```
{s₁[t] + s₂[t] + s₃[t] + s₄[t]} == {Const[1]}
```

If we now modify this metabolic system by assuming that the concentrations of S_1 and S_4 are fixed, as we did in the question/answer in Section 4.5, the conservation relationships disappear, as is indicated by the null matrix, denoted by {}, below.

```
S̄ = Table[Sᵢ[t], {i, 2, 3}];
```

$$\tilde{N} = \begin{pmatrix} 1 & -1 & 0 \\ 0 & 1 & -1 \end{pmatrix};$$

```
ConservationRelations[S̄, Ñ, GMatrix → True]
```

```
{}
```

Q: Calculate the conservation relationships and the vector of steady-state concentrations for the following reaction scheme that is written in the *Mathematica* Cell:

```
Clear[Const];

eqns = {{s₁[t]  →  s₂[t]},
        {s₂[t] + s₅[t]  →  s₃[t] + s₆[t] },
```

```
                {s₃[t] → s₄[t]},
                {s₆[t] → s₅[t]}};

        Ñ = NMatrix[eqns, {s₁[t], s₄[t]}];
        S̄ = SMatrix[eqns, {s₁[t], s₄[t]}];

        v[1]  = 1;
        v[2]  = s₂[t] s₅[t];
        v[3]  = s₃[t];
        v[4]  = s₆[t];

        v̄ := VMatrix[eqns, {s₁[t], s₄[t]}];
```

A: The conservation relations of this system are obtained by using the function **ConservationRelations** from the **MetabolicControlAnalysis** package.

```
        ConservationRelations[S̄, Ñ]
```

```
    {s₅[t] + s₆[t]} == {Const[1]}
```

Hence, the conservation relationship is that the sum of the concentrations of S_5 and S_6 is constant.

If we now determine the steady-state concentrations of the system, it is seen that the result is given in terms of the constant, **Const[1]**.

```
        steadystate = SteadyState[S̄, Ñ, v̄]
```

$$\left\{\left\{s_2[t] \to \frac{1}{-1 + \text{Const}[1]},\right.\right.$$
$$\left.\left. s_3[t] \to 1, s_5[t] \to -1 + \text{Const}[1], s_6[t] \to 1\right\}\right\}$$

These constants can be assigned values like other parameters. For example,

```
        Const[1] = 1.5;
        steadystate
```

$$\{\{s_2[t] \to 2., s_3[t] \to 1, s_5[t] \to 0.5, s_6[t] \to 1\}\}$$

Thus the concentrations of metabolites at the steady state are defined in terms of both the metabolite-conservation relationships and the kinetic parameters.

From an examination of the simple examples in this section, the reader may be wondering about the utility of the **ConservationRelations** function, given the rather obvious form of the conservation relations that we have determined thus far. This function is important for two reasons: (1) it is a mathematical necessity to be able to specify these conservation relations so that we can solve for the steady-state concentrations. As seen in the previous question/answer, the solution of the steady-state vector will be expressed in terms of the constants of these relationships; (2) in models of much more complicated metabolic systems, these relationships can become less obvious. Hence the function **ConservationRelations** becomes particularly important. We will return to these issues again in Section 7.5 where we deal with determining steady states and conservation relationships in a realistic cellular system, namely, the metabolism in the human erythrocyte.

4.7 Stability of a Steady State

After the calculation of the steady-state concentrations of metabolites in a reaction scheme it is often relevant to determine if the particular steady state is stable. A steady state is said to be *asymptotically* stable if, after a minor perturbation in system concentrations, the system returns to the original set of values. In other words, as the time since perterbation approaches infinity, the system will approach the original steady state. If the perterbation causes the system to move toward another steady state, or to not reach a steady state at all, then the steady state is said to be *unstable*. In the **MetabolicControlAnalysis** package the function **Stability** determines whether a set of simultaneous differential equations, defined by $\mathbf{S}$, $\mathbf{N}$, $\mathbf{v}$, and $\mathbf{p}$, is stable or not at a given steady state. It can be shown mathematically that the presence or absence of asymptotic stability is determined by the so-called eigenvalues of the Jacobian of the system of differential equations.[1] The Jacobian is defined as follows[1] by

$$\text{Jacobian}(\mathbf{M}) = \mathbf{N}\frac{\partial\,\mathbf{v}}{\partial\,\mathbf{S}} \text{ or } m_{ij} = \sum_{k}\frac{\partial v_k}{\partial S_j}\ . \qquad [4.11]$$

If the eigenvalues all have negative real parts, the steady state is said to be asymptotically stable. If at least one of the eigenvalues has a positive real part, the steady state is *unstable*. In the case that some of the eigenvalues have zero real parts (with the remaining eigenvalues all having negative real parts) no conclusions can been drawn about the stability. In this final case, further analysis is needed (see Heinrich and Schuster[1] and references therein).

Stability[*S*, *N*, *v*, *p*, SteadyStateConc → *steadystate*]	Assesses whether the system of differential equations, defined by *S*, *N*, *v*, and *p* is asymptotically stable at the steady state given by the replacement rule *steadystate*. Also, the function returns the *eigenvalues* of the Jacobian of the system of differential equations defined by *S*, *N*, *v*, and *p*. Note that the inclusion of the parameter table *p* is optional
MMatrix[*S*, *N*, *v*, *p*, Normalized → False, SteadyStateConc → *steadystate*]	Calculates the Jacobian of the system of differential equations defined by *S*, *N*, *v*, and *p*. Note that the last three arguments are optional; however, the default value for the option Normalized is True

Assessing the stability of a steady state.

Q: Assess the stability of the steady state calculated in the question/answer in Section 4.5.

A: The function **Stability** requires the full description of the reaction scheme, and in addition it must have the full set of steady-state concentrations. Thus, the reaction list is

```
Clear[Subscript, Const]
eqns = { {s₁[t]  →  s₂[t]},
         {s₂[t]  →  s₃[t]},
         {s₃[t]  →  s₄[t]}};
```

The S, N, and v matrices and the parameters are defined as they were previously in Section 4.6.

```
S̄ := SMatrix[eqns, {s₁[t] , s₄[t]}] ;
Ñ = NMatrix[eqns, {s₁[t] , s₄[t]}] ;
```

$$Do\left[v[i] = \frac{V_{max,i,f}\, s_i[t]}{K_{m,i,s_i}\left(1 + \frac{s_i[t]}{K_{m,i,s_i}}\right)}, \{i, 3\}\right];$$

```
v̄ := VMatrix[eqns] ;
```

```
p = {s₁[t], s₄[t], Vmax,1,f, Vmax,2,f, Vmax,3,f, Km,1,s₁, Km,2,s₂, Km,3,s₃};
pv = {1, 1, 1, 1, 1, 1, 1, 1};
```

If we determine the steady-state vector, we can then use the **Stability** function to assess the stability of the particular steady state.

```
ss = SteadyState[S̄, Ñ, v̄, Transpose[{p, pv}]]
```

```
Stability[S̄, Ñ, v̄, Transpose[{p, pv}], SteadyStateConc → ss₍₁₎]
```

```
{{s₂[t] → 1, s₃[t] → 1}}
```

```
Asymptotically Stable
```

$$\left\{ -\frac{1}{4}, -\frac{1}{4} \right\}$$

This output verifies that the steady state is stable, since the eigenvalues returned by the **Stability** function are all real and negative.

Q: Calculate the Jacobian of the reaction system

$$S_1 \xrightarrow{v_1} S_2 \xrightarrow{v_2} \, , \qquad [4.12]$$

assuming that both reactions are each dependent on S_1 and S_2. Thus, $v_1 \equiv v_1(S_1, S_2)$ and $v_2 \equiv v_2(S_1, S_2)$.

A: The following series of functions defines the reaction scheme and then **MMatrix** carries out the evaluation of the Jacobian.

```
Clear[v];

S̄ = Table[s₁[t], {i, 1, 2}];
Ñ = IdentityMatrix[2];
v̄ := Table[v[i][S̄], {i, 1, 2}];

jacob = MMatrix[S̄, Ñ, v̄, Normalized -> False]

{{v[1]^({1,0}) [{s₁[t], s₂[t]}], v[1]^({0,1}) [{s₁[t], s₂[t]}]},
 {v[2]^({1,0}) [{s₁[t], s₂[t]}], v[2]^({0,1}) [{s₁[t], s₂[t]}]}}
```

The eigenvalues of the Jacobian are determined by using the *Mathematica* function **Eigenvalues**. The sign of the real part of these eigenvalues must be negative for the system to be stable at the given steady state.

```
Eigenvalues[jacob]
```

$$\{ \frac{1}{2} \left(v[2]^{(\{0,1\})} [\{s_1[t], s_2[t]\}] + v[1]^{(\{1,0\})} [\{s_1[t], s_2[t]\}] - \right.$$

$$\sqrt{\left(v[2]^{(\{0,1\})} [\{s_1[t], s_2[t]\}]^2 - 2\, v[2]^{(\{0,1\})} [\{s_1[t], s_2[t]\}] \right.}$$

$$v[1]^{(\{1,0\})} [\{s_1[t], s_2[t]\}] + v[1]^{(\{1,0\})} [\{s_1[t], s_2[t]\}]^2 +$$

$$\left. \left. 4\, v[1]^{(\{0,1\})} [\{s_1[t], s_2[t]\}]\, v[2]^{(\{1,0\})} [\{s_1[t], s_2[t]\}] \right) \right),$$

$$\frac{1}{2} \left(v[2]^{(\{0,1\})} [\{s_1[t], s_2[t]\}] + v[1]^{(\{1,0\})} [\{s_1[t], s_2[t]\}] + \right.$$

$$\sqrt{\left(v[2]^{(\{0,1\})} [\{s_1[t], s_2[t]\}]^2 - 2\, v[2]^{(\{0,1\})} [\{s_1[t], s_2[t]\}] \right.}$$

$$v[1]^{(\{1,0\})} [\{s_1[t], s_2[t]\}] + v[1]^{(\{1,0\})} [\{s_1[t], s_2[t]\}]^2 +$$

$$\left. \left. \left. 4\, v[1]^{(\{0,1\})} [\{s_1[t], s_2[t]\}]\, v[2]^{(\{1,0\})} [\{s_1[t], s_2[t]\}] \right) \right) \}$$

Clearly, the sign of the eigenvalues is dependent on the expressions for the velocities and the values of their parameters, as well as the actual steady-state values themselves.

Q: What is the steady-state concentration of S_1 in the reaction scheme defined below (from Heinrich and Shuster,[1] Scheme 4, p. 47)? The scheme involves the supply of the metabolite, S_1, and two pathways of removal, with one displaying high-substrate inhibition of the enzyme that catalyzes it.

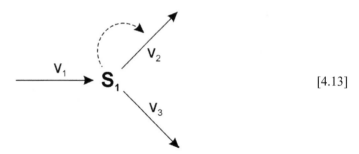

[4.13]

Suppose that $v_1 = k_1$, $v_2 = k_2[S_1]$, and $v_3 = \frac{k_3[S_1]}{1+(\frac{[S_1]}{K_i})^n}$, with $k_1 = 0.9$ mmol L^{-1} h^{-1}, $k_2 = 0.11$ h^{-1}, $k_3 = 0.4$ h^{-1}, $K_i = 3$ mmol L^{-1}, and $n = 4$.

A: First we need to define the appropriate matrices, reaction rates, and parameter values for this system.

```
Clear[Subscript];

S̃ = {s₁[t]};

Ñ = {{1, -1, -1}};

v[1] := k₁;
v[2] := k₂ s₁[t];
```

$$v[3] := \frac{k_3 \, s_1[t]}{1 + \left(\frac{s_1[t]}{K_i}\right)^n} \, ;$$

```
v̄ := Table[v[j], {j, 3}];

p = {k₁, k₂, k₃, Kᵢ, n};
pv = {0.9, 0.11, 0.4, 3, 4};
```

Then the steady state is calculated as follows:

```
ss = SteadyState[S̄, N̄, v̄, Transpose[{p, pv}]]
```

$\{\{s_1[t] \to -2.33223 - 2.7614 \, i\}, \{s_1[t] \to -2.33223 + 2.7614 \, i\},$
$\{s_1[t] \to 2.05637\}, \{s_1[t] \to 3.28846\}, \{s_1[t] \to 7.50144\}\}$

By selecting only the real solutions we find three physically relevant steady states.

```
realss = ss /. {a___, b_ -> c_Complex, d___} -> {} // Flatten
```

$\{s_1[t] \to 2.05637, s_1[t] \to 3.28846, s_1[t] \to 7.50144\}$

We can now use a **Do** statement around the **Stability** function to assess the stability of each steady state individually.

```
Do[Stability[S̄, N̄, v̄, Transpose[{p, pv}],
   SteadyStateConc -> realss⟦i⟧] , {i, 3}]
```

Asymptotically Stable

{-0.200649}

Asymptotically Unstable

{0.113127}

Asymptotically Stable

{-0.0810647}

The stability analysis reveals that the second steady state in the list is asymptotically unstable. This is illustrated in the graph from the following simulation where the initial concentration of S_1 is set to the second value. (More significant figures are used than are shown above.)

```
sol = NDSolveMatrix[S̄, Ñ, v̄,
    {s₁[0] == 3.28846091596}, {t, 0, 800}, Transpose[{p, pv}]];

Plot[Evaluate[S̄ /. sol], {t, 0, 800},
    AxesLabel -> {"Time (h)", "Concentration (mM)"},
    PlotRange → {1, 4}];
```

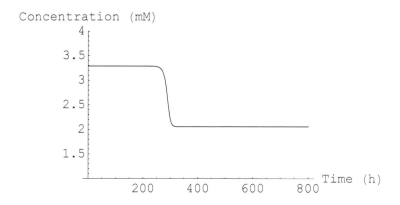

Figure 4.5. Time course of the reaction scheme shown in Eqn [4.13].

The graph shows that the concentration is stable at ~3.3 mmol L^{-1} and after ~320 h it undergoes a transition to a new steady state at ~2.0 mmol L^{-1}.

4.8 When Cell Volume Changes with Time

All the metabolic pathways that we have considered so far have involved the assumption that the reactions take place in free solution and no consideration has been given to partitioning of reactants and enzymes into various compartments. To actually simulate such a system introduces another order of complexity into formulating the model. The principal 'trick' in such models is to express the rate equations in terms of the rate of change of amounts (moles) rather than concentrations. If we do this Eqn [4.3] becomes

$$\frac{d(\mathbf{V.S})}{dt} = \mathbf{N v} \ , \qquad\qquad [4.14]$$

where $\mathbf{N}$ is the stoichiometry matrix, $\mathbf{v}$ is the vector of reaction velocities containing rate equations expressed in terms of the time rate of change of amounts, $\mathbf{S}$ is the vector of substrate concentrations, and $\mathbf{V}$ is a diagonal matrix where the ith element contains the compartment volume of the ith substrate in **S**. Note that in this formulation, although the rate equations are expressed in terms of the rate of change of amounts, the rate equations are also expressed as functions of substrate concentrations. In other words, all we do for this analysis is to take the usual rate expressions and multiply them by the appropriate compartment volumes.

If we expand Eqn [4.14] we obtain

$$\mathbf{V}\frac{d\mathbf{S}}{dt} + \mathbf{S}\frac{d\mathbf{V}}{dt} = \mathbf{N}\mathbf{v} . \qquad [4.15]$$

Now for many systems of interest we can assume that the compartment volumes remain constant with time. In this case Eqn [4.15] simplifies to

$$\mathbf{V}\frac{d\mathbf{S}}{dt} = \mathbf{N}\mathbf{v} , \qquad [4.16]$$

or

$$\frac{d\mathbf{S}}{dt} = \mathbf{V}^{-1}\mathbf{N}\mathbf{v} . \qquad [4.17]$$

This may seem a little abstract at the moment but it should become clearer with the following question/answer.

Q: Consider the following simple two-compartment model (Eqn [4.18]) consisting of an intracellular and an extracellular compartment with volumes Vol_i and Vol_e, respectively.

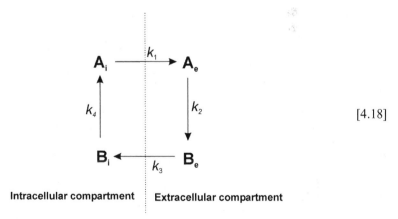

[4.18]

Intracellular compartment | **Extracellular compartment**

The model contains only two metabolites A and B; the subscript e denotes that the metabolite is in the extracellular compartment, while the subscript i refers to the intracellular compartment. Assume that all reactions are described by first-order rate constants. Simulate a time course for this system and determine the steady-state concentrations of A and B in each compartment.

A: The reaction scheme represented in Eqn [4.18] can be described by the following equation list:

```
eqns = {
    {Aᵢ[t]  →  Aₑ[t]},
    {Aₑ[t]  →  Bₑ[t]},
    {Bₑ[t]  →  Bᵢ[t]},
    { Bᵢ[t]  →  Aᵢ[t]}};
```

We model each of these transport and metabolic reactions with simple first-order rate equations as indicated in the question. Note that because we want to express the reaction/transport rates in terms of the number of molecules per unit time, it is necessary to multiply each rate equation by the appropriate compartment volume. Suppose that the compartment volumes are

```
Volᵢ = 0.75;
Volₑ = 0.25;
```

Hence, the requisite rate equations are

```
v[1]  := k₁ Aᵢ[t] Volᵢ;              k₁ = 1;

v[2]  := k₂ Aₑ[t] Volₑ;              k₂ = 1;

v[3]  := k₃ Bₑ[t] Volₑ;              k₃ = 1;

v[4]  := k₄ Bᵢ[t] Volᵢ;              k₄ = 1;
```

Now use the **MetabolicControlAnalysis** functions to derive the matrices and vectors of this system, in the manner used above.

```
S̄ = SMatrix[eqns];
N̄ = NMatrix[eqns];
v̄ = VMatrix[eqns];
S̄ // MatrixForm
N̄ // MatrixForm
v̄ // MatrixForm
```

$$\begin{pmatrix} A_e[t] \\ A_i[t] \\ B_e[t] \\ B_i[t] \end{pmatrix}$$

$$\begin{pmatrix} 1 & -1 & 0 & 0 \\ -1 & 0 & 0 & 1 \\ 0 & 1 & -1 & 0 \\ 0 & 0 & 1 & -1 \end{pmatrix}$$

$$\begin{pmatrix} 0.75\,A_i\,[t] \\ 0.25\,A_e\,[t] \\ 0.25\,B_e\,[t] \\ 0.75\,B_i\,[t] \end{pmatrix}$$

It remains to define the **V** matrix of Eqn [4.14]; this is simply the diagonal matrix where the ith element contains the compartment volume of the ith substrate in **S**. Hence for this system, **V** is given by

```
V̄ = DiagonalMatrix[{Vole, Voli, Vole, Voli}];
V̄ // MatrixForm
```

$$\begin{pmatrix} 0.25 & 0 & 0 & 0 \\ 0 & 0.75 & 0 & 0 \\ 0 & 0 & 0.25 & 0 \\ 0 & 0 & 0 & 0.75 \end{pmatrix}$$

When constructing this matrix, make sure that the order of the volumes in the diagonal matrix matches the order of the substrates in the substrate list.

Having constructed all the appropriate matrices we can solve this system of matrices and vectors by using the **NDSolveMatrix** function in the manner described above, except that we use $\mathbf{V}^{-1}\,\mathbf{N}$ instead of simply **N**.

```
sol = NDSolveMatrix[S̄, Inverse[V̄].N̄, v̄,
  {Ai[0] == 1, Ae[0] == 0, Bi[0] == 0, Be[0] == 0}, {t, 0, 10}]

{{Ae → InterpolatingFunction[{{0., 10.}}, <>],
  Ai → InterpolatingFunction[{{0., 10.}}, <>],
  Be → InterpolatingFunction[{{0., 10.}}, <>],
  Bi → InterpolatingFunction[{{0., 10.}}, <>]}}

Plot[Evaluate[S̄ /. sol], {t, 0, 10}, PlotRange -> {0, 1.5},
  AxesLabel -> {"Time", "Concentration"}];
```

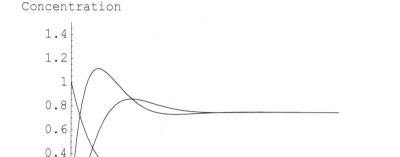

Figure 4.6. Time course of the reaction scheme shown in Eqn [4.18]. A_i delines from an initial concentration of 1 to a steady state of 0.25 mmol L^{-1}, while B_i approaches a steady state of 0.25 mmol L^{-1} from an initial concentration of 0 mmol L^{-1}. Both $[A_e]$ and $[B_e]$ approach a steady state of 0.75 mmol L^{-1} with $[A_e]$ initially overshooting this steady-state concentration by a modest amount.

From this simulated time course it is seen that the system approaches a steady state. To determine the values of the concentrations we use the function **SteadyState**; however, we must first determine the conservation of mass relations; these are given by the following function from the MetabolicControlAnalysis package.

ConservationRelations[$\bar{S}$, Inverse[$\bar{V}$].$\bar{N}$]

> $\{A_e[t] + 3. A_i[t] + 1. B_e[t] + 3. B_i[t]\} == \{Const[1]\}$

From the initial conditions used in the time course simulation, we must set **Const[1] = 3**.

Const[1] = 3;

Hence, the steady state of the system is given by the following function evaluation:

SteadyState[$\bar{S}$, Inverse[$\bar{V}$].$\bar{N}$, $\bar{v}$]

$\{\{A_e[t] \to 0.75, A_i[t] \to 0.25, B_e[t] \to 0.75, B_i[t] \to 0.25\}\}$

In the above example we assumed that cell volumes were constant. In many situations of interest this may not be the case. For example, most cells are unable to sustain a high transmembrane difference in osmotic pressure and, if this occurs, the cell volume will change.

To simulate metabolic systems in which compartment volumes change we could resort to using Eqn [4.15]. However, another method, which often makes model formulation easier, is to represent the changes as changes in amounts rather than concentrations. When doing so the familiar equations are still employed:

$$\frac{d\mathbf{S}}{dt} = \mathbf{N}\mathbf{v} \; . \tag{4.19}$$

However, in this case, $\mathbf{S}$, is a list of substrate amounts (not concentrations), while the rate equations in $\mathbf{v}$ are written in terms of the rate of change of amounts. The stoichiometric matrix, $\mathbf{N}$, remains unchanged. This overall approach to modelling these systems is illustrated in the following question/answer.

Q: Simulate the time dependence of the concentrations of Na, Ca, and K in a suspension of red blood cells in which a calcium ionophore is added to the cells. The ionophore stimulates K efflux by delivering Ca into the cells, thus 'switching on' the calcium-activated (Gardos) K channel.

A: In this problem it is assumed that a cell is not able to sustain a substantial difference of osmolality across its membrane. Hence, as ions flow across the membrane, water distribution is altered and this is reflected in a change in the cell volume. Furthermore, since the metal cations permeate the cell membrane, but charged intracellular proteins do not, a potential difference is generated across the membrane. This is called the *Donnan potential*. Its value is calculated by using the Nernst equation (see below). A graphical illustration of the reaction scheme is as follows:

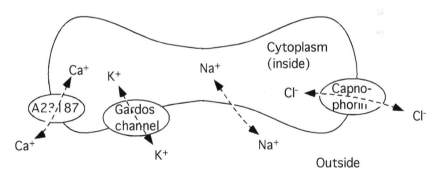

Figure 4.7. Model of a suspension of cells that take up calcium that, in turn, activates the loss of potassium from the cells. The calcium influx is also accompanied by exchange of sodium between the inside and the outside of the cells. The scheme is a model of human erythrocytes in which calcium activates the Gardos channel. The model consists of a Michaelis-Menten membrane transport protein for each of the cationic species, while it is assumed, as is the case with human erythrocytes, that the passive exchange of anions (chloride and bicarbonate) is very rapid (on the sub-second timescale) via the band 3 (capnophorin) transport protein.

The first step in modelling this system is to define carefully all the initial volumes in the system. Suppose that the total sample volume is 3×10^{-3} L. Hence we define the values as follows:

`vol`_{total} `= 3.0 × 10`⁻³`;`

We now specify the fraction of the sample volume that is occupied by the cells. This fraction is called the hematocrit. Note that the hematocrit potentially changes in time as the ions exchange across the cell membrane and an input value is the haematocrit at zero time. Thus,

`ht[0] = 0.65;`

The volume fraction of an erythrocyte that is occupied by haemoglobin and the cytoskeleton is usually taken to be 0.29 of the volume of the cell of normal volume. When the cell changes volume, the volume of these proteins stays the same but their volume *relative* to the cell volume changes. Thus the fraction of the erythrocyte volume that is occupied by haemoglobin is given by

`vol`_{hbn} `= ht[0] × 0.29 × vol`_{total}`;`

Now use these values to calculate the initial water volumes in the intracellular ($vol_{i,w}[0]$) and extracellular ($vol_e[0]$) space of the sample.

```
vole [0] = (1 - ht[0]) × voltotal;
voli [0] = ht[0] × voltotal;
voli,w [0] = voli [0] - volhbn;
voltotal,w [0] = voli,w [0] + vole [0];
```

Having defined the initial water volumes in the two compartments of the sample, we must determine how these two volumes change with time. Before we do this it is necessary to define the ion movements that will lead to these volume changes.

The system of transport processes that is described in Fig. 4.7 can be represented with the following equation list:

```
eqns = {{kitrans,    ki [t]    →    ke [t] },
        {ketrans,    ke [t]    →    ki [t] },
        {naitrans,   nai [t]   →    nae [t] },
        {naetrans,   nae [t]   →    nai [t] },
        {caitrans,   cai [t]   →    cae [t] },
        {caetrans,   cae [t]   →    cai [t] }};
```

Each of the cationic transport processes is modelled by a Michaelis-Menten rate equation. For simplicity we assume that intracellular calcium activates the influx and efflux of both sodium and potassium ions. The rate equations and their associated parameters are as follows:

$\mathbf{V}_{\max,ki} = 0.0061 \, \mathbf{vol}_i[0]$;

$\mathbf{K}_{m,ki} = 2.5 \times 10^{-4}$;

$\mathbf{K}_{ac} = 2.5 \times 10^{-5}$;

$$\mathbf{v[kitrans]} := \frac{\mathbf{V}_{\max,ki} \, \frac{k_i[t]}{vol_{i,w}[t]}}{\frac{k_i[t]}{vol_{i,w}[t]} + \mathbf{K}_{m,ki} \left(1 + \frac{K_{ac} \, vol_{i,w}[t]}{ca_i[t]}\right)} ;$$

$\mathbf{V}_{\max,ke} = 0.0061 \, \mathbf{vol}_i[0]$;

$\mathbf{K}_{m,ke} = 2.5 \times 10^{-4}$;

$$\mathbf{v[ketrans]} := \frac{\mathbf{V}_{\max,ke} \, \frac{k_e[t]}{vol_e[t]}}{\frac{k_e[t]}{vol_e[t]} + \mathbf{K}_{m,ke} \left(1 + \frac{K_{ac} \, vol_{i,w}[t]}{ca_i[t]}\right)} ;$$

$\mathbf{V}_{\max,nai} = 0.00045 \, \mathbf{vol}_i[0]$;

$\mathbf{K}_{m,nai} = 2.5 \times 10^{-3}$;

$$\mathbf{v[naitrans]} := \frac{\mathbf{V}_{\max,nai} \, \frac{na_i[t]}{vol_{i,w}[t]}}{\frac{na_i[t]}{vol_{i,w}[t]} + \mathbf{K}_{m,nai} \left(1 + \frac{K_{ac} \, vol_{i,w}[t]}{ca_i[t]}\right)} ;$$

$\mathbf{V}_{\max,nae} = 0.00045 \, \mathbf{vol}_i[0]$;

$\mathbf{K}_{m,nae} = 2.5 \times 10^{-3}$;

$$\mathbf{v[naetrans]} := \frac{\mathbf{V}_{\max,nae} \, \frac{na_e[t]}{vol_e[t]}}{\frac{na_e[t]}{vol_e[t]} + \mathbf{K}_{m,nae} \left(1 + \frac{K_{ac} \, vol_{i,w}[t]}{ca_i[t]}\right)} ;$$

$\mathbf{V}_{\max,cai} = 0.0122 \, \mathbf{vol}_i[0]$;

$\mathbf{K}_{m,cai} = 2.0 \times 10^{-6}$;

$$\mathbf{v[caitrans]} := \frac{\mathbf{V}_{\max,cai} \, \frac{ca_i[t]}{vol_{i,w}[t]}}{\frac{ca_i[t]}{vol_{i,w}[t]} + \mathbf{K}_{m,cai}} ;$$

$\mathbf{V}_{\max,cae} = 0.0122 \, \mathbf{vol}_i[0]$;

$K_{m,cae} = 2.0 \times 10^{-6};$

$$v[caetrans] := \dfrac{V_{max,cae} \dfrac{ca_e[t]}{vol_e[t]}}{\dfrac{ca_e[t]}{vol_e[t]} + K_{m,cae}};$$

Note that each rate equation is specified in terms of the rate of change of amount (units: mol s^{-1}) rather than the rate of change of concentration. Because of this the V_{max} parameters in the above equations have units of mol s^{-1} and are made up as follows: number of molecules of transporter per cell (N) × turnover number per active site of the transporter × number of cells in the sample. The total number of cells in the sample is given by vol_{total} × ht[0] / volume of a single cell.

Hence, when determining V_{max} values from the literature, we must pay special attention to its physical units. The V_{max} values we needed for the above equations are reported in terms of mol (L internal volume)$^{-1}$ s^{-1} so it is necessary to convert them to the appropriate units by multiplying by the internal volume of the cell.

Having defined the systems of reactions and the associated rate equations, we are now able to define the **S**, **N**, and **v** matrices, and the initial conditions.

```
S̄ = SMatrix[eqns];
N̄ = NMatrix[eqns];
v̄ = VMatrix[eqns];

IC = {nae[0] == 0.14 × vole[0],
nai[0] == 0.005 × voli,w[0],
ke[0] == 0.010 × vole[0],
ki[0] == 0.14 × voli,w[0],
cae[0] == 0.002 × vole[0],
cai[0] == 0.0001 × voli,w[0]};
```

The final task before we can simulate a timecourse is to define how cell volume will change with changes in ionic concentrations. Our model of volume change requires the following major assumption: that the passive exchange of anions is rapid compared to the rates of transport of other ions. With this assumption we know that chloride ions (and other anions such as bicarbonate) will distribute between the intracellular and extracellular spaces so that electroneutrality is maintained in each compartment. Thus the following equation must hold:

```
cle[t_] := nae[t] + ke[t] + 2 cae[t];
cli[t_] := nai[t] + ki[t] + 2 cai[t] - 4 hb;
```

Note that we have assumed that the charge on each haemoglobin molecule is − 4.

In addition we know that the erythrocyte membrane is not able to withstand an osmotic pressure difference of any great magnitude. So it is appropriate to assume that the osmotic pressure inside the cell is equal to the osmotic pressure outside the cell. This assumption is equivalent to assuming that the sum of the concentrations of all solutes inside is equal to the sum of all those outside, under conditions of thermodynamic 'ideality'. Hence, the following equation will yield the intracellular volume as a function of the concentrations of the ionic species inside and outside the cell:

```
voli,w[t_] :=
    (voltotal,w[0] (ki[t] + nai[t] + cai[t] + cli[t] + hb)) / (ki[t] +
        nai[t] + cai[t] + cli[t] + hb + ke[t] + nae[t] + cae[t] + cle[t]);
```

where the mass (mol) of haemoglobin, hb, is calculated by using its partial specific volume (0.73 L kg^{-1}) and its molecular weight (64.5 kg).

$$hb = \frac{vol_{hbn}}{0.73 \times 64.5};$$

The volume of aqueous medium outside the cells is simply the total solute-accessible volume in the sample minus the aqueous volume inside the cells.

```
vole[t_] := voltotal,w[0] - voli,w[t];
```

We are now in a position to solve the time-dependent behaviour of this system using **NDSolveMatrix** and then to view the time courses of concentrations or amounts graphically.

```
res = NDSolveMatrix[S̄, Ñ, v̄, IC, {t, 0, 5000}, MaxSteps -> 5000];


Plot[Evaluate[ki[t] /. res],
    {t, 0.0, 5000} , AxesLabel → {"Time (s)", "K⁺ (mol)"},
    PlotRange → {0.0, 0.0002}];
```

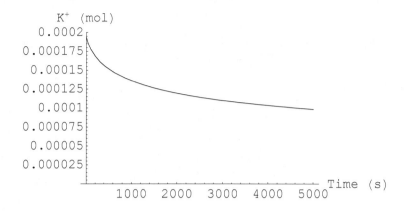

Figure 4.8. Simulated time course of the amount of intracellular potassium ion in a cell suspension for the reaction scheme shown in Fig. 4.7.

```
Plot[Evaluate[{ca_i[t] / vol_i,w[t], ca_e[t] / vol_e[t]} /. res],
    {t, 160, 320}, PlotRange → All,
    AxesLabel → {"Time (s)", "[Ca⁺]_in and [Ca⁺]_out (mol L⁻¹)"}];
```

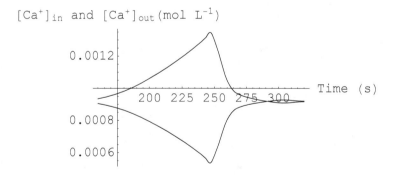

Figure 4.9. Simulated time course of the concentration of intra- (lower curve) and extracellular calcium ions for the reaction scheme shown in Fig. 4.7.

```
Plot[Evaluate[{vol_i,w[t] + vol_hbn, vol_e[t]} /. res],
    {t, 0, 5000}, PlotRange → All,
    AxesLabel → {"Time (s)", "Vol_in and Vol_out (L)"}];
```

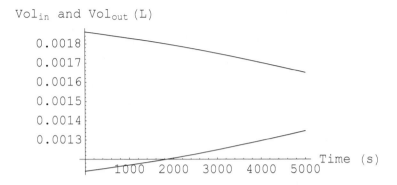

Figure 4.10. Simulated time course of the intra- (upper curve) and extracellular volumes for the reaction scheme shown in Fig. 4.7.

Note that the volumes do not reach the same value, primarily because the hemoglobin and other proteins inside the cells occupy volume.

From these results we can also calculate the membrane, based on the anion distribution ratio, by using the Nernst equation as follows:

```
rR = 8.314; tT = 310; fF = 96500;
parNernst = rR tT / fF;

gph4 = Plot[1000 parNernst * Log[Evaluate[cl_i[t] / cl_e[t] /. res]],
    {t, 0, 5000}, PlotRange → All,
    AxesLabel → {"Time (s)", "Membrane Potential(mV)"}];
```

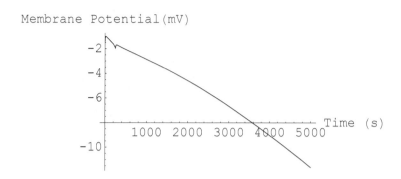

Figure 4.11. Simulated time course of the membrane potential for the reaction scheme shown in Fig. 4.7.

4.9 Decompostion of N and Calculation of the Link Matrix (Optional)

The functions **NDSolveMatrix**, **SteadyState**, and **Conservation-Relations**, as well as the add-on functions given in Chapter 5 which are functions in the add-on package **MetabolicControlAnalysis**, all involve a rearrangement of the stoichiometry matrix. This rearrangement is usually hidden from the user. However, by using the command **LinkMatrix**, information can be obtained on what rearrangements have been made in **N** to expedite the solution of the matrix form of the rate equations. The link matrix is given this name because it contains stoichiometric information that links various reactants as co-substrates in various reactions in the metabolic scheme.

LinkMatrix[**S**, **N**] Rearranges the rows of
N so that its upper rank **N** rows are
linearly independent and form a submatrix
N^o. Also returns a 'Link Matrix, ' **L**,
such that $N = L N^o = N^o$

Transforming the matrix **N** into one that is suitable for use in metabolic calculations.

Q: Determine the link matrix for the reaction scheme given in the previous question/answer in Section 4.6.

A: The stoichiometry matrix is re-entered and the function **LinkMatrix** performs the analysis.

```
Clear[Const];

eqns = {{S[1][t] → S[2][t]},
        {S[2][t] + S[5][t] → S[3][t] + S[6][t] },
        {S[3][t] → S[4][t]},
        {S[6][t] → S[5][t]}};

Ñ = NMatrix[eqns, {S[1][t], S[4][t]}];
S̄ = SMatrix[eqns, {S[1][t], S[4][t]}];

linkoutput = LinkMatrix[S̄, Ñ]
```

$\{\{S̄, \{S[2][t], S[3][t], S[6][t], S[5][t]\}\},$
$\{L, \{\{1, 0, 0\}, \{0, 1, 0\}, \{0, 0, 1\}, \{0, 0, -1\}\}\},$
$\{Ñ^0, \{\{1, -1, 0, 0\}, \{0, 1, -1, 0\}, \{0, 1, 0, -1\}\}\},$
$\{Ñ, \{\{1, -1, 0, 0\}, \{0, 1, -1, 0\}, \{0, 1, 0, -1\}, \{0, -1, 0, 1\}\}\},$
$\{G, \{\{0, 0, 1, 1\}\}\}, \{LinkMatrixTransform, \{\{3, 4\}\}\}\}$

Thus the function **LinkMatrix** returns a table, the first element of which contains the rearranged **S** vector. The second element is **L**, the third **N⁰**, and the fourth the rearranged **N**. The fifth element contains the matrix **G** which has the property[1]

$$\mathbf{GN} = \mathbf{0} , \qquad\qquad [4.20]$$

while the last element contains rules for transforming the old **N** into the rearranged **N**. Each transformation rule is given as a two-element list $\{a, b\}$ indicating that row a has been moved to row b.

Note that **G** returned by **LinkMatrix** is always in the form $(-\mathbf{L}\ \mathbf{I})$. This is not always the case for **G** calculated with the function **ConservationRelations**. The latter **G** is calculated in a manner that ensures that the conservation relationships are not negative.

The merit of using these transformed matrices lies in some of the deeper aspects of MCA that are not developed further in this book.

4.10 Excercises

4.10.1

Use the function **NMatrix** from the **MetabolicControlAnalysis** add-on package to determine and interpret the stoichiometric matrix for the reaction scheme given in Eqn [4.4].

4.10.2

Numerically determine the steady-state vector of concentrations for the reaction scheme given in Section 4.5; however, use an initial estimate of 0.0 mmol L^{-1} for both $[S_2]$ and $[S_3]$.

4.10.3

Determine the conservation relations for the reaction scheme given in Eqns [4.5] – [4.8]. Give a cell-physiological interpretation for each of the derived relations.

4.10.4

What can be said about the stability of the steady state determined in the last worked question/answer in Section 4.6?

4.11 References

1. Heinrich, R. and Schuster, S. (1996) *The Regulation of Cellular Systems*. Chapman & Hall, New York.

5 Metabolic Control Analysis

5.1 Introduction

In Chapter 4 we used *Mathematica* to simulate the time-dependent behavior of some simplistic metabolic systems and to determine the existence of stable or unstable steady-state modes. These types of analyses are extremely important as they allow us to make predictions with a model and to ask a myriad of "what if?" questions about the system. The tests that were used are amongst the first steps in testing a mathematical model against experimental results. These steps are vital for assessing the validity or predictive usefulness of a model. Indeed we use these techniques again when we study parameter estimation and model refinement in Chapter 6. These steps are also the first to be applied before MCA is applied to a model of a metabolic scheme.

MCA is an analysis paradigm in the field of metabolism that was developed to give rigorous, quantitative answers to questions such as "What are the roles of the individual reactions in determining selected metabolite concentrations and fluxes through different segments of the pathway?" The major concepts of MCA were formulated in the 1970s by Kascer and Burns[1] and Heinrich and Rapoport.[2] The ideas had earlier become well known to engineers who referred to them as *sensitivity analysis*. The analysis enables the determination of the sensitivity of a response or output of a dynamical system to variations in the value of a selected parameter in the system.

In this chapter *Mathematica* is used to perform MCA on mathematical models of metabolism. Only an introduction to MCA is given; however, sufficient information is provided to allow sophisticated MCA on complicated and realistic metabolic systems. More detailed reviews of the theoretical and experimental aspects of MCA appear in several texts and journal articles.[3-6]

This chapter also introduces more aspects of our add-on package, **MetabolicControlAnalysis**. It includes functions that are useful for performing MCA on systems of differential equations that are represented in vector and matrix form. The matrix notation and various analytical procedures that are used in these functions are those presented by Heinrich and Schuster;[5] and readers who are interested in the theory behind the calculations are encouraged to consult this excellent book.

Finally, we will illustrate the ideas of MCA by using some simple examples. Because of the use of matrix algebra, there are times when one may feel that the mathematics is an unnecessarily complicated way to express relatively simple ideas. However, bear with the presentation; the utility of the method of expressing the ideas mathematically in matrix form will become more apparent when we apply these methods to the biochemically important and realistic examples in Chapters 7 and 8.

5.2 Control Coefficients

A major focus of MCA is the quantification of the role that individual enzymic reactions play in determining metabolite concentrations and pathway fluxes at a particular steady state of metabolite concentrations. Fundamental to the theory is the definition and calculation of *elasticity* coefficients, *response* coefficients, and *control* coefficients. These three types of coefficient are dealt with, in turn, in this and the following few sections. It is also important to note that MCA was developed primarily for the analysis of systems under steady-state conditions. Hence all the coefficients are defined for a particular reference steady state.

The description of the MCA coefficients begins with the one that, historically, was the first to be defined: for a metabolic system of i metabolites, j metabolite fluxes, and k reactions, at a particular steady state, the *flux control coefficients* (FCC) are defined as

$$C_{v_k}^{J_j} = \frac{v_k}{J_j} \frac{\partial J_j / \partial p_k}{\partial v_k / \partial p_k} \quad , \quad\quad\quad [5.1]$$

where p_k is a parameter that only affects the velocity of reaction k (v_k), and $\partial v_k / \partial p_k$ is evaluated with the parameter values and concentrations associated with the particular steady state. Note that the change in flux is normalized by dividing it by the reference value at the steady state, and the change in reaction velocity is also normalized by dividing it by the value at the steady state. Thus the FCC expresses quantitatively the effect that varying the parameter p_k has on the flux through the system, J_j, if the effect of p_k on the local enzyme rate, v_k, is known. In other words, the FCC is a measure of the extent to which the intrinsic rate of reaction k controls the steady-state flux J.

```
<<                        Reads in the addon package
MetabolicControlAnalysi·.  MetabolicControlAnalysis.
s`

 FluxControlMatrix[S, N,   Calculates a matrix for the metabolic
 v, p, SteadyState → steadystate]  system defined by the substrate vector,
                          S, the stoichiometric matrix, N,
                          and the parameter matrix, p, at the steady
                          state given by the replacement rule steadystate;
                          where the element mᵢⱼ is the flux control coefficient
                          of the flux through reaction i with respect to
                          reaction j. Note that the last two arguments are optional.
```

Calculating flux control coefficients.

Q: Consider the example from Hofmeyer and Cornish-Bowden;[7] a metabolite, M, is produced at a rate, v_s, by a supply (source) pathway and consumed at a rate, v_d, by a demand (sink) pathway.

$$\xrightarrow{v_s} M \xrightarrow{v_d} \quad\quad [5.2]$$

Assuming that both v_s and v_d are influenced by M, what are the FCC values for pathway flux with respect to the supply and demand pathways?

A: First we load the add-on package **MetabolicControlAnalysis** and define the reactant vector **S**, the stoichiometry matrix **N**, and the vector of velocity or rate expressions **v**, for this system.

```
<< MetabolicControlAnalysis`;

S̄ = {M[t]};
Ñ = ( 1  -1 );
v̄ := {v[s][M[t]], v[d][M[t]]};
```

By assuming that in the steady state $M[t] = M_{ss}$, we apply the function **FluxControlMatrix** to return a matrix of flux control coefficients.

```
FluxControlMatrix[S̄, Ñ, v̄,
  SteadyStateConc → {M[t] -> Mₛₛ}] // MatrixForm
```

$$\begin{pmatrix} 1 - \dfrac{v[s]'[M_{ss}]}{-v[d]'[M_{ss}]+v[s]'[M_{ss}]} & \dfrac{v[d][M_{ss}]\,v[s]'[M_{ss}]}{v[s][M_{ss}]\,(-v[d]'[M_{ss}]+v[s]'[M_{ss}])} \\ -\dfrac{v[s][M_{ss}]\,v[d]'[M_{ss}]}{v[d][M_{ss}]\,(-v[d]'[M_{ss}]+v[s]'[M_{ss}])} & 1 + \dfrac{v[d]'[M_{ss}]}{-v[d]'[M_{ss}]+v[s]'[M_{ss}]} \end{pmatrix}$$

From the elements in row 2 column 1, and row 1 column 2, we can write the following expressions for the flux control coefficients:

$$C_{v_s}^J = \frac{\varepsilon_M^{v_d}}{\varepsilon_M^{v_d} - \varepsilon_M^{v_s}} \text{ and } C_{v_d}^J = \frac{-\varepsilon_M^{v_s}}{\varepsilon_M^{v_d} - \varepsilon_M^{v_s}} \text{ ,} \qquad [5.3]$$

where $\varepsilon_M^{v_d} = v[d]'[M_{ss}]/v[d][M_{ss}]$ and $\varepsilon_M^{v_s} = v[s]'[M_{ss}]/v[s][M_{ss}]$. Note that these expressions have been simplified by using the fact that in the steady-state $v_s = v_d$. Be careful to look closely at the meanings of these two expressions: the numerator specifies the dependence of a change in reaction rate with respect to [M] when [M] is equal to the steady-state concentration, and the denominator expresses the velocity of the reaction at the steady-state concentration of M. These terms, $\varepsilon_M^{v_d}$ and $\varepsilon_M^{v_s}$, are called elasticity coefficients (see Section 5.3).

An important point regarding this question/answer is that $C_{v_s}^J + C_{v_d}^J = 1$. This is an example of the *summation theorem* for flux control coefficients.[1–3,5,6] Formally, the theorem states that the sum of all flux control coefficients in a metabolic system for a particular metabolic flux, J_j, is equal to 1, i.e.,

$$\sum_k C_{v_k}^{J_j} = 1 \text{ .} \qquad [5.4]$$

Hence, the entries along each row of the matrix of flux control coefficients should add up to 1. This relationship is only true for normalized flux control coefficients, namely, flux control coefficients that are defined according to Eqn [5.1].

In a manner similar to the flux control coefficient, a *concentration control coefficient* is defined as

$$C_{v_k}^{S_i} = \frac{v_k}{S_i} \frac{\partial S_i / \partial p_k}{\partial v_k / \partial p_k} \qquad [5.5]$$

where S_i is the concentration of the metabolite i. Notice that the expression contains the normalization of reaction rate and metabolite concentration with respect to the corresponding steady-state value of S_i. (See Eqn [5.9] below for a relaxation of this condition.)

ConcControlMatrix[S, N,	Returns a matrix where the element
v, p, SteadyState → *steadystate*]	m_{ij} is the concentration control coefficient
	of metabolite i with respect to reaction j.

Calculating concentration control coefficients.

Q: Consider the reaction scheme shown in the previous question/answer. Calculate the concentration control coefficient for M in this scheme.

A: We begin the solution by specifying **S**, **N**, and **v** in the same form as they were used above, and then we implement the function **ConcControlMatrix** from the **MetabolicControlAnalysis** package (which we have already loaded).

```
ConcControlMatrix[S̄, Ñ, v̄,
    SteadyStateConc → {M[t] -> Mₛₛ}] // MatrixForm
```

$$\left(-\frac{v[s][M_{ss}]}{M_{ss}\,(-v[d]'[M_{ss}]+v[s]'[M_{ss}])} \quad \frac{v[d][M_{ss}]}{M_{ss}\,(-v[d]'[M_{ss}]+v[s]'[M_{ss}])} \right)$$

Using the same definitions for ε as above, we obtain

$$C_{v_s}^M = \frac{1}{\varepsilon_M^{v_d} - \varepsilon_M^{v_s}} \quad \text{and} \quad C_{v_d}^M = \frac{-1}{\varepsilon_M^{v_d} - \varepsilon_M^{v_s}} \ . \qquad [5.6]$$

In a manner like flux control coefficients, there is also a summation theorem for concentration control coefficients. It is

$$\sum_k C_{v_k}^{S_i} = 0 \ . \qquad [5.7]$$

Hence entries along each row of the matrix of concentration control coefficients add up to 0. This is illustrated in the previous question/answer where $C_{v_s}^M + C_{v_d}^M = 0$.

Q: Consider the reaction scheme below in Eqn [5.8] that is characterized by the system of differential equations: $v[1] = k_1\, s_1[t]$, $v[2] = k_2\, s_2[t]$, and $v[3] = k_3\, s_3[t]$.

$$S_1 \xrightarrow{v_1} S_2 \xrightarrow{v_2} S_3 \xrightarrow{v_3} \qquad [5.8]$$

Calculate the concentration control matrix for this system.

A: A suitable program to solve this problem is

```
S̄ = {s₂[t], s₃[t]};

Ñ = ( 1  -1   0  );
    ( 0  +1  -1 )

v[1] := k₁ s₁[t];
v[2] := k₂ s₂[t];
v[3] := k₃ s₃[t];

v̄ := Table[v[i], {i, 3}];

(*Parameter vector.*)
p = {k₁, k₂, k₃, s₁[t]};
pv = {1, 2, 1, 1};
p̄ = Transpose[{p, pv}];
```

Before we can calculate the matrix of concentration control coefficients, we must calculate the metabolite concentrations at the steady state. This is achieved with the **SteadyState** function from the **MetabolicControlAnalysis** add-on package.

```
solution = SteadyState[S̄, Ñ, v̄, p̄]
```

$$\left\{\left\{s_2[t] \to \frac{1}{2}, s_3[t] \to 1\right\}\right\}$$

There is only one steady state for this reaction scheme, and the matrix of concentration control coefficients for this steady state is given by using the **ConcControlMatrix** function, that is also from the **MetabolicControlAnalysis** add-on package.

```
ConcControlMatrix[S̄, Ñ, v̄, p̄,
    SteadyStateConc -> solution⟦1⟧] // MatrixForm
```

$$\begin{pmatrix} 1 & -1 & 0 \\ 1 & 0 & -1 \end{pmatrix}$$

From this matrix it can be seen that the concentration of S_2 depends only on reactions 1 and 2 (elements 1 an -1 in the first two columns of the first row, respectively), while the concentration of S_3 depends only on reactions 1 and 3 (elements 1 and -1 in the first and third columns of the second row, respectively).

Q: Consider again the reaction scheme in Eqn [5.8]. What happens to the concentration control coefficient for S_2 when reaction v_1 becomes subject to feedback inhibition by S_3?

A: The feedback inhibition can be modelled by modifying the equation for v_1 by including an inhibition term that is akin to the competitive inhibition factor in a Michaelis-Menten equation (Section 2.2.2). Hence we suppose that the inhibition of v_1 is described by the equation $v[1] = k_1 s_1[t]/(1+s_3[t]/K_{i,1})$. Then we redefine the rate vector and parameter vector from the previous question/answer as follows:

```
v[1] := k₁ s₁[t] ;
        ─────────
        1 + s₃[t]
            ────
            Kᵢ,₁

v̄ := Table[v[j], {j, 3}];


p = {k₁, k₂, k₃, Kᵢ,₁, s₁[t]};
pv = {1, 2, 1, 0.3, 1};
p̄ = Transpose[{p, pv}];
```

The steady state for this new metabolic system is obtained by using the function **SteadyState** from the **MetabolicControlAnalysis** add-on package.

```
solution = SteadyState[S̄, N̄, v̄, p̄]

{{s₂[t] → -0.358945, s₃[t] → -0.717891},
 {s₂[t] → 0.208945, s₃[t] → 0.417891}}
```

Of the two pairs of lists in **solution** only one is physically meaningful, namely, the one having non-negative values of the steady state concentrations of S_2 and S_3. These realistic concentrations are next used via the replacement rule, **SteadyStateConc -> solution[[2]]**, in the function **ConcControlMatrix** that calculates the matrix of concentration control coefficients.

```
ConcControlMatrix[S̄, N̄, v̄, p̄,
    SteadyStateConc -> solution[[2]]] // MatrixForm

( 0.632068  -1.   0.367932  )
( 0.632068   0.  -0.632068 )
```

The main conclusion that can be drawn from this matrix is that the reaction v_3 that consumes the metabolite S_3 that inhibits reaction 1 now shares some of the control over the concentration of S_2. Specifically, the element in the third column of the first row is now non-zero, being 0.367932.

The functions **FluxControlMatrix** and **ConcControlMatrix** also have the option of returning non-normalized control coefficients. Non-normalized flux and concentration control coefficients are defined by

$$C_{v_k}^{J_j} = \frac{\partial J_j / \partial p_k}{\partial v_k / \partial p_k} \quad \text{and} \quad C_{v_k}^{S_i} = \frac{\partial S_i / \partial p_k}{\partial v_k / \partial p_k} , \qquad [5.9]$$

respectively, and can be calculated from the functions by including the option **Normalized → False**.

Q: Calculate non-normalized control coefficients for the reaction scheme analyzed in the previous question/answer.

A: For this operation we apply the **ConcControlMatrix** function as previously but this time we add the option **Normalized → False**, as follows:

ConcControlMatrix[S̄, N̄, v̄, p̄, SteadyStateConc -> solution₍₂₎,
   Normalized -> False] // MatrixForm

$$\begin{pmatrix} 0.316034 & -0.5 & 0.183966 \\ 0.632068 & 0. & -0.632068 \end{pmatrix}$$

One notable feature of this normalized matrix is that the elements in each row sum to 0; this is expected according to the corresponding summation theorem (Eqn [5.7]).

5.3 Calculation of Control Coefficients by Numerical Perturbation

The functions presented in the previous section, in the **MetabolicControl-Analysis** add-on package, allow control coefficients to be conveniently and rapidly calculated. The algorithms used in these functions are given in Appendix 2 and are based on the matrix methods described by Heinrich and Schuster.[5] However, for most realistic metabolic systems, application of these methods can lead to significant numerical errors. (For a discussion of this point see Section 8.4.) Hence it is useful to have another more generally applicable method for estimating control coefficients.

This alternative method of calculating control coefficients is to use numerical perturbation. This involves replacing the partial differentials in the above equations (Eqns [5.1], [5.5], [5.9]) by finite differences. For example, we can numerically approximate the differential, $\partial J_j / \partial p_k$, in Eqn [5.1] by making a small (say, 0.1%) change in the value or the parameter and then resimulating a time course to determine system fluxes. $\partial J_j / \partial p_k$ will then be approximated by the difference in the flux values before and after perturbation divided by 0.001. This technique is illustrated in the following example.

Q: For the reaction scheme given in Eqn [5.8], calculate the value of $C_{v_1}^{S_2}$ by numerical perturbation.

A: A parameter that we can perturb in order to affect v_1 is k_1. Hence we must set up two parameter tables, one containing the original parameter values and the other exactly the same except that the value of k_1 is increased by 0.1% of its original value. This is done as follows:

```
p = {k₁, k₂, k₃, Kᵢ,₁, s₁[t]};
pv1 = {1, 2, 1, 0.3, 1}; (*Vector of parameter values.*)
pv2 = {1 * 1.001, 2, 1, 0.3, 1}; (*k[1] multiplied by 1.001.*)

p̄1 = Transpose[{p, pv1}]; (*Parameter table.*)
p̄2 = Transpose[{p, pv2}];
(*Second parameter table with k[1] increased by 0.1%*)
```

The effect of the perturbation of the parameter k_1 on the steady-state concentration of S_2 can now be determined by evaluating steady states using the original and then the perturbed parameter tables. This is done with the function **SteadyState** in the **MetabolicControlAnalysis** add-on package.

```
solution1 = SteadyState[s̄, Ñ, v̄, p̄1];
solution2 = SteadyState[s̄, Ñ, v̄, p̄2];

s₂,₁ = s₂[t] /. solution1⟦2⟧ ;
s₂,₂ = s₂[t] /. solution2⟦2⟧ ;
```

$$\text{concControlCoeff} = \frac{s_{2,2} - s_{2,1}}{s_{2,1}} \times \frac{1}{0.001}$$

```
0.631921
```

The value of 0.631921 of the concentration control coefficient compares well with the value of 0.632068 calculated previously by the matrix method.

An important point to note when applying the numerical perturbation method for calculating control coefficients is that for many enzymic reactions the following relationship holds:

$$\frac{e_k}{v_k} \frac{\partial v_k}{\partial e_k} = 1 \ , \tag{5.10}$$

and hence Eqns [5.1] and [5.5] reduce to

$$C_{v_k}^{J_j} = \frac{e_k}{J_j} \frac{\partial J_j}{\partial e_k} \quad \text{and} \quad C_{v_k}^{S_i} = \frac{e_k}{S_i} \frac{\partial S_i}{\partial e_k} \ , \tag{5.11}$$

respectively. Thus, for many reactions control coefficients can be estimated simply by perturbing the total enzyme concentration. For an example of this see Section 8.4.

5.4 Elasticity Coefficients

Another derived parameter in MCA is the elasticity coefficient. This coefficient is different from the previous two in that elasticity coefficients are properties of individual enzymes in isolation, rather than of the metabolic system as a whole. Essentially, elasticity coefficients measure how the velocity of an enzymic reaction changes in response to changes in substrate, product, inhibitor, and activator concentrations. Hence, these coefficients quantify what enzyme kineticists have been measuring since the days of Michaelis and Menten.

More formally, an elasticity coefficient characterizes the response of an individual enzyme in isolation to the perturbation of a parameter or metabolite concentration; and it is determined at a reference state of substrate, product, and effector concentrations. Elasticity coefficients are of two types, being defined as follows:

$$\varepsilon_{S_j}^{v_i} = \frac{S_j}{v_i} \frac{\partial v_i}{\partial S_j} \; ; \varepsilon - \text{elasticities} \quad, \qquad [5.12]$$

$$\pi_{p_k}^{v_i} = \frac{p_k}{v_i} \frac{\partial v_i}{\partial p_k} \; ; \pi - \text{elasticities} \quad, \qquad [5.13]$$

where v_i is the velocity of reaction i, S_j is the concentration of metabolite S_j and p_k is a parameter that affects v_i. Note that ε-elasticities and π-elasticities are essentially the same thing, mathematically.   The reason to distinguish them apart is that computationally it is necessary to treat the metabolite concentrations (state variables) differently from the model or so-called *structural* parameters like rate constants and total enzyme concentrations.

EpsilonElasticityMatrix[$S, N, v, p,$ SteadyState $\rightarrow$ *steadystate*]	Returns a matrix where the element m_{ij} is the $\varepsilon -$ elasticity of reaction i with respect to substrate j.
PiElasticityMatrix[$S, N, v, \{parameter list\}, p,$ SteadyState $\rightarrow$ *steadystate*]	Returns a matrix where the element m_{ik} is the $\pi -$ elasticity of reaction i with respect to the kth parameter of the parameter list.

Calculating elasticity coefficients.

Q: For the modified reaction scheme of Eqn [5.8] ($S_1 \xrightarrow{v_1} S_2 \xrightarrow{v_2} S_3 \xrightarrow{v_3}$; v[1] = $k_1 \, s_1[t]/(1+s_3[t]/K_{i,1})$, v[2] = $k_2 \, s_2[t]$, and v[3] = $k_3 \, s_3[t]$), calculate the matrices of (1) ε-elasticity coefficients; and (2) π-elasticity coefficients for all parameters.

A: (1) This is achieved with the following function from the **MetabolicControl-Analysis** add-on package.

```
EpsilonElasticityMatrix[S̄, Ñ, v̄, p̄,
   SteadyStateConc -> solution₍₂₎] // MatrixForm
```

$$\begin{pmatrix} 0. & -0.582109 \\ 1. & 0. \\ 0. & 1. \end{pmatrix}$$

Recall from Eqn [5.12] that the rows refer to reactions $(1-3)$ and the columns to the substrates (S_2 and S_3); so the top left-hand element indicates that v_1 is unaffected by S_2, but the ε-elasticity coefficient (top right-hand element of the matrix) is non-zero and negative. This is consistent with the imposed *negative feedback* or inhibition by S_3 of v_1. From the second row it is seen that the elasticity of v_2 with respect to S_2 is 1 but the elasticity of S_3 with respect to this reaction is 0; this is predicted on the grounds that the rate equation for v_2 has no term in S_3 implying that S_3 does not influence the rate of the reaction. A similar argument can be made for the interpretation of the third row of the matrix.

(2) The function **PiElasticityMatrix** is also in the **MetabolicControl-Analysis** add-on package. Thus, the π-elasticity matrix is calculated in an analogous manner to the ε-elasticity matrix except that an additional argument {parameter list} must be included, as follows:

```
p = {k₁, k₂, k₃, K_{i,1}, s₁[t]};

PiElasticityMatrix[S̄, Ñ, v̄, p, p̄,
   SteadyStateConc -> solution⟦2⟧] // MatrixForm
```

$$\begin{pmatrix} 1. & 0. & 0. & 0.582109 & 1. \\ 0. & 1. & 0. & 0. & 0. \\ 0. & 0. & 1. & 0. & 0. \end{pmatrix}$$

Recall that each column corresponds to a parameter in the reaction scheme, in the order specified by the **p** vector; while the rows correspond to the reactions as specified by the **v̄** vector. When applying **PiElasticityMatrix** only those parameters for which elasticity coefficients are to be calculated need to be included in the parameter list or **p** vector.

For the functions **EpsilonElasticityMatrix** and **PiElasticityMatrix** there is the option of returning non-normalized coefficients; these are defined by

$$\varepsilon^{v_i}_{S_j} = \frac{\partial v_i}{\partial S_j} \; ; \varepsilon - \text{elasticities} \;, \qquad [5.14]$$

$$\pi^{v_i}_{p_k} = \frac{\partial v_i}{\partial S_j} \; ; \pi - \text{elasticities} \;, \qquad [5.15]$$

and they are calculated by including the option **Normalized → False**.

Q: For the modified reaction scheme of Eqn [5.8] ($S_1 \xrightarrow{v_1} S_2 \xrightarrow{v_2} S_3 \xrightarrow{v_3}$; v[1] = k_1 s_1[t]/(1+s_3[t]/$K_{i,1}$), v[2] = k_2 s_2[t], and v[3] = k_3 s_3[t]), calculate the matrix of non-normalized π-elasticity coefficients for all parameters.

A: This is done as in the previous question/answer except that the additional argument **Normalized → False** is included.

> **PiElasticityMatrix[S̄, Ñ, v̄, p, p̄, SteadyStateConc -> solution$_{[\![2]\!]}$,**
> **Normalized -> False] // MatrixForm**

$$\begin{pmatrix} 0.417891 & 0 & 0 & 0.81086 & 0.417891 \\ 0 & 0.208945 & 0 & 0 & 0 \\ 0 & 0 & 0.417891 & 0 & 0 \end{pmatrix}$$

The non-normalized matrix has zeros in the same positions as for the normalized case above, but the non-zero elements have substantially different values from the previous matrix. The significance of this is discussed later.

Q: Calculate the matrix of non-normalized ε-elasticity coefficients of reaction 2 in the two metabolite system shown in the scheme in Eqn [5.16]. Each enzymic reaction has a dependence of its rate on the concentrations of S_1 and S_2 that is represented by $v_1(S_1, S_2)$ and $v_2(S_1, S_2)$.

$$S_1 \xrightarrow{v_1} S_2 \xrightarrow{v_2} \qquad\qquad [5.16]$$

A: The calculation is begun by setting up a Table of values of S_1 and S_2 denoted by **S̄**; a stoichiometry matrix, **Ñ**, that in this case is the identity matrix; and a vector, **v̄**, of rate equations. Then the function **EpsilonElasticityMatrix** from the **MetabolicControlAnalysis** add-on package is used to generate the matrix of elasticity coefficients.

> **Clear[v];**
> **S̄ = Table[s$_i$[t], {i, 1, 2}];**
> **Ñ = IdentityMatrix[2];** (*2 x 2 identity matrix.*)
> **v̄ := Table[v[i][S̄], {i, 1, 2}];**
>
> **EpsilonElasticityMatrix[S̄, Ñ, v̄, Normalized -> False] //**
> **MatrixForm**
>
> $$\begin{pmatrix} v[1]^{(\{1,0\})}[\{s_1[t], s_2[t]\}] & v[1]^{(\{0,1\})}[\{s_1[t], s_2[t]\}] \\ v[2]^{(\{1,0\})}[\{s_1[t], s_2[t]\}] & v[2]^{(\{0,1\})}[\{s_1[t], s_2[t]\}] \end{pmatrix}$$

Thus, the element in row 1 column 1 is the derivative of v_1 with respect to S_1, while the element in row 1 column 2 is the derivative of v_1 with respect to S_2. The second row contains derivatives of v_2.

5.5 Response Coefficients

The final type of coefficients of major importance in MCA are the *response coefficients*. These characterize the effect of an infinitesimally small change in a parameter value in the system on concentrations or fluxes in the system of reactions. Thus, the concentration response coefficient is defined by

$$R^{S_i}_{p_k} = \frac{p_k}{S_i} \frac{\partial S_i}{\partial p_k} \,,$$ [5.17]

and the flux response coefficient is defined by

$$R^{J_j}_{p_k} = \frac{p_k}{S_j} \frac{\partial J_j}{\partial p_k} \,,$$ [5.18]

ConcResponseMatrix[**S, N, v,** {*paramter list*}, **p,** SteadyState → *steadystate*]	Returns a matrix where the element m_{ik} is the concentration response coefficient of the concentration of metabolite i with respect to the kth parameter of parameter list.
FluxResponseMatrix[**S, N, v,** {*paramter list*}, **p,** SteadyState → *steadystate*]	Returns a matrix where the element m_{jk} is the flux response coefficient of the flux through reaction j with respect to the kth parameter of parameter list.

Calculating response coefficients.

Q: (1) For the modified reaction scheme of Eqn [5.8] ($S_1 \xrightarrow{v_1} S_2 \xrightarrow{v_2} S_3 \xrightarrow{v_3}$; v[1] = $k_1 \, s_1[t]/(1+s_3[t]/K_{i,1})$, v[2] = $k_2 \, s_2[t]$, and v[3] = $k_3 \, s_3[t]$) calculate the concentration response matrix for all parameters. (2) What would be the expected response in [S_3] if the parameter k_2 were changed?

A: (1) This problem is solved by defining the metabolite vector, $\bar{\mathbf{S}}$, the stoichiometry matrix, $\bar{\mathbf{N}}$, the three rate expressions that are placed in the vector, $\mathbf{v}$, the parameter vector, **p**, and the vector of corresponding parameter values, **pv**. Then a function-call to **ConcResponseMatrix**, from the **MetabolicControlAnalysis** add-on package, yields the solution.

$\bar{S} = \{s_2[t], s_3[t]\};$

$$\bar{N} = \begin{pmatrix} 1 & -1 & 0 \\ 0 & +1 & -1 \end{pmatrix};$$

$$v[1] := \frac{k_1 s_1[t]}{1 + \frac{s_3[t]}{K_{i,1}}};$$

$v[2] := k_2 s_2[t];$

$v[3] := k_3 s_3[t];$

$\bar{v} := \text{Table}[v[j], \{j, 3\}];$

$p = \{k_1, k_2, k_3, K_{i,1}, s_1[t]\};$

$pv = \{1, 2, 1, 0.3, 1\};$

$\bar{p} = \text{Transpose}[\{p, pv\}];$

$\text{ConcResponseMatrix}[\bar{S}, \bar{N}, \bar{v}, p, \bar{p},$

$\quad \text{SteadyStateConc} \rightarrow \text{solution}_{[\![2]\!]}] \,// \,\text{MatrixForm}$

$$\begin{pmatrix} 0.632068 & -1. & 0.367932 & 0.367932 & 0.632068 \\ 0.632068 & 0. & -0.632068 & 0.367932 & 0.632068 \end{pmatrix}$$

The matrix represents the concentration response coefficients with those pertaining to S_2 (in the column order specified in the vector **p**) in the first row and to S_3 in the second row. While not intending to discuss all 10 coefficients, it is useful to note that the concentration response coefficient of S_3 with respect to $v[2] = k_2\, s_2[t]$ is 0 (second row and second column of the matrix). This outcome is as expected simply on the grounds that the expression for $v[2]$ has no term in S_3 and hence the derivative of $v[2]$ with respect to S_3 is 0. Or, physically, the reaction characterized by v_2 does not have any mechanistic involvement of S_3.

(2) Hence the answer to the second part of the question is that the entry in row 2 column 2 gives the response coefficient for S_3 with respect to the parameter k_2. Since this entry is 0, a perturbation of k_2 would be expected to have no effect on [S_3] under all conditions of reactant concentrations.

From MCA and the definition of the control coefficients it follows that response coefficients can be written in terms of control coefficients and π-elasticities. This outcome is as follows:

$$R^{S_i}_{p_k} = \sum_j C^{S_i}_{v_j} \pi^{v_j}_{p_k} \,,$$

$$[5.19]$$

$$\text{and} \quad R^{J_j}_{p_k} = \sum_i C^{J_j}_{v_i} \pi^{v_i}_{p_k} \quad .$$

[5.20]

Thus by definition the key MCA functions in the **MetabolicControlAnalysis** add-on package have the following interrelationship:

ConcResponseMatrix = ConcControlMatrix . PiElasticityMatrix
and
FluxResponseMatrix = FluxControlMatrix . PiElasticityMatrix.

Q: The matrix of concentration response coefficients given in the previous question/answer can also be calculated by using the following relationship between the key matrices:

ConcResponseMatrix = ConcControlMatrix · PiElasticityMatrix.

Verify this.

A: The *Mathematica* implementation of this analysis for the reaction scheme used in the previous question/answer is

```
ConcControlMatrix[S̄, Ñ, v̄, p̄, SteadyStateConc -> solution[[2]]] .
   PiElasticityMatrix[S̄, Ñ, v̄, p, p̄,
   SteadyStateConc -> solution[[2]]] // MatrixForm
```

$$\begin{pmatrix} 0.632068 & -1. & 0.367932 & 0.367932 & 0.632068 \\ 0.632068 & 0. & -0.632068 & 0.367932 & 0.632068 \end{pmatrix}$$

This gives the same result as previously obtained.

From Eqns [5.19] and [5.20] it is seen that the total response from the perturbation of a parameter is the sum of the individual responses from each reaction. These individual responses $C^{S_i}_{v_j} \pi^{v_j}_{p_k}$ or $C^{J_j}_{v_i} \pi^{v_i}_{p_k}$ are defined as the *partial response coefficients*.

PartialConcResponse[	Returns a matrix of partial
S, N, v, n, {parameter list},	concentration response coefficients for a
p, SteadyState → *steadystate*]	metabolite at position *n* in *S*. Each
	entry m_{jk} in the matrix gives the
	product of the concentration control coefficient
	with respect to reaction *j* and the π-elasticity
	coefficient with respect to reaction *j* and parameter *k*.
PartialFluxResponse[	Returns a matrix of
S, N, v, n, {parameter list},	partial flux response coefficients for a
p, SteadyState → *steadystate*]	flux at position *n* in *v*. Each entry
	m_{jk} in the matrix gives the
	product of the flux control coefficient with
	respect to reaction *j* and the π-elasticity
	coefficient with respect to reaction *j* and parameter *k*.

Calculating partial response coefficients.

Q: For the reaction scheme given in Eqn [5.8] calculate the matrix of partial concentration response coefficients for S_2.

A: This is achieved with the function **PartialConcResponse** from the **MetabolicControlAnalysis** add-on package.

> **PartialConcResponse[$\bar{S}$, $\tilde{N}$, $\bar{v}$, 1, p,**
> **$\bar{p}$, SteadyStateConc -> solution$_{[2]}$] // MatrixForm**

$$\begin{pmatrix} 0.632068 & 0. & 0. & 0.367932 & 0.632068 \\ 0. & -1. & 0. & 0. & 0. \\ 0. & 0. & 0.367932 & 0. & 0. \end{pmatrix}$$

The interpretation of this matrix is as follows. Each column corresponds to the consecutive members of the parameter set $\{k_1, k_2, k_3, K_{i,1}, s_1[t]\}$ and each row to the reactions v_1, v_2, and v_3. Each column in the matrix describes how the total response to a parameter perturbation is partitioned amongst the reactions of the system. Hence the sum of entries in each column will add up to the total response coefficient.

For example, from the concentration response matrix calculated in the previous question/answer, the response coefficient of $[S_2]$ to k_1 is 0.632. From column 1 in the partial concentration response matrix for $[S_2]$ calculated above, it is seen that this response is entirely due to reaction 1. Although this result is hardly surprising, this type of analysis can be very useful in more complicated metabolic networks (see Chapter 8).

5.6 Internal Response Coefficients

Two more coefficients that are useful for describing the regulation of metabolic pathways are the internal response coefficients. These are defined in a manner that is analogous to the response coefficients but they use ε-elasticities instead of π-elasticities. The definition of the response that characterizes the effects of a fluctuation in concentration of S_j on the concentration of S_i is

$$\tilde{R}_{S_j}^{S_i} = \sum_k^N C_{v_k}^{S_i} \varepsilon_{S_j}^{v_k} = \sum_k^N {}^k R_{v_k}^{S_i} \quad , \tag{5.21}$$

and for the dependence of the flux J_i on the concentration of S_j the definition of the *internal response coefficient* is

$$R_{S_j}^{J_i} = \sum_k^N C_{v_k}^{J_i} \varepsilon_{S_j}^{v_k} = \sum_k^N {}^k R_{v_k}^{S_i} \quad . \tag{5.22}$$

Q: Calculate the internal response matrices for the reaction scheme given in Eqn [5.8].

A: (1) The internal concentration response matrix is calculated as follows:

```
ConcControlMatrix[S̄, Ñ, v̄, p̄, SteadyStateConc -> solution⟦2⟧].
   EpsilonElasticityMatrix[S̄, Ñ, v̄, p̄,
     SteadyStateConc -> solution⟦2⟧] // MatrixForm
```

$$\begin{pmatrix} -1. & 0. \\ 0. & -1. \end{pmatrix}$$

The interpretation of the output matrix is that S_2 has a negative effect on its own concentration (top left-hand element of the matrix), and this is unsurprising since it is the substrate for the reaction that removes itself. On the other hand, for S_3, by construction of the reaction scheme and the rate equations, there is no mechanistic action of S_3 on the concentration of S_2 so the top right-hand element of the matrix is 0. The same argument applies to the effect of S_2 on S_3 and hence the relevant values in the second row of the matrix.

(2) For the second part of the question, the following product returns the internal flux response matrix.

```
FluxControlMatrix[S̄, Ñ, v̄, p̄, SteadyStateConc -> solution⟦2⟧].
   EpsilonElasticityMatrix[S̄, Ñ, v̄, p̄,
     SteadyStateConc -> solution⟦2⟧]  // MatrixForm
```

$$\begin{pmatrix} 0. & 0. \\ 0. & 0. \\ 0. & 5.55112 \times 10^{-17} \end{pmatrix}$$

The interpretation of this matrix is as follows. Each column corresponds to the consecutive members of the reactant set { S_2, S_3 } and each row to the v_1, v_2, and v_3. A column in the matrix describes how a flux through each reaction in the linear sequence is partitioned amongst the reactions of the system. Hence the sum of entries in each column will add up to the total response coefficient of the system. In this open reaction system the elements of the matrix are all zero, thus implying that the fluxes through each reaction are not controlled by either S_2 acting via S_3 or vice versa for S_3 acting via S_2. While this outcome is easy to visualize and understand for a simple system such as that in Eqn [5.8], such may not be true for complicated models of metabolism as discussed in Chapter 8.

The results in the last question/answer are instances of the general theoretical result relating to the internal response coefficients, that

$$\tilde{R}_{S_j}^{S_i} = \sum_{k}^{N} C_{v_k}^{S_i}\, \varepsilon_{S_j}^{v_k} = \sum_{k}^{N} {}^k R_{S_j}^{S_i} = -\delta_{ij} \quad , \tag{5.23}$$

$$\text{and} \quad \tilde{R}_{S_j}^{J_i} = \sum_{k}^{N} C_{v_k}^{J_i}\, \varepsilon_{S_j}^{v_k} = \sum_{k}^{N} {}^k R_{S_j}^{J_i} = 0 \quad . \tag{5.24}$$

Each ${}^k R_{S_j}^{S_i}$ or ${}^k R_{S_j}^{J_i}$ is called the *partial internal response coefficient*. The coefficient is a measure of the contribution of the kth reaction to the total response of the system. From Eqn [5.23] we obtain the following result for the summation of the coefficients:

$$-\tilde{R}_{S_i}^{S_i} = -\sum_{k} {}^k R_{S_i}^{S_i} = 1 \quad , \tag{5.25}$$

where each $-{}^k R_{S_i}^{S_i} \equiv {}^k H_{S_i}^{S_i}$ is a measure of the contribution of each kth reaction to the restoration of homeostasis after a perturbation in S_i. ${}^k H_{S_i}^{S_i}$ is called the *homeostatic strength* of the kth reaction with respect to metabolite S_i.

PartialInternalConcResp·. **onse[**S, N, v, n, m, p, SteadyState → *steadystate***]**	Returns a vector which contains the partial internal concentration response coefficients for a metabolite at position n in S with respect to a metabolite at position m in S. The jth position in the vector is the partial internal response coefficient for reaction j.
PartialInternalFluxResp·. **onse[**S, N, v, n, m, p, SteadyState → *steadystate***]**	Returns a vector which contains the partial internal flux response coefficients for a flux at position n in v with respect to a metabolite at position m in S. The jth position in the vector is the partial internal response coefficient for reaction j.

Calculating partial internal response coefficients.

Q: For the reaction scheme given in Eqn [5.8] calculate the homeostatic strength of all reactions with respect to S_3.

A: This calculation is done with the function **PartialInternalConcResponse** from the **MetabolicControlAnalysis** add-on package. (Note that the homeostatic strength is the negative of the partial internal concentration response coefficient.)

```
- PartialInternalConcResponse[S̄, Ñ, v̄, 2, 2,
  p̄, SteadyStateConc -> solution[2]] // MatrixForm
```

$$\begin{pmatrix} 0.367932 \\ 0 \\ 0.632068 \end{pmatrix}$$

This result indicates that the consumption of S_3 by reaction 3 is more important than the feedback inhibition of reaction 1 by S_3 for maintaining homeostasis of S_3. This is a remarkable insight and certainly one that can be fruitfully explored by varying the relative values of the kinetic parameters. Such excursions are left for you, the reader.

5.7 Conclusions

This chapter has introduced the various control coefficients that exist in the modern theory of MCA.[5] Time and space have not permitted extensive examples of their use and interpretation in this chapter. However, this is used extensively in Chapters 7 and 8.

On the other hand, it is worth summarizing the fact that all the MCA coefficients for any simulatable pathway can be determined by using the relevant function from the **MetabolicControlAnalysis** add-on package that is presented here. The list of coefficients and their corresponding package functions is as follows: (1) concentration control coefficient, **ConcControlMatrix**; (2) flux control coefficient, **Flux-ControlMatrix**; (3) reactant-concentration- or ε-elasticity, **EpsilonElas-ticityMatrix**; (4) parameter-value- or π-elasticity, **PiElasticityMatrix**; (5) concentration response coefficient, **ConcResponseMatrix**; (6) flux response coefficient, **FluxResponseMatrix**; (7) partial internal response coefficient with respect to the concentration of a reactant, **PartialInternalConcResponse**; and (8) and the partial internal response coefficient with respect to a flux, **Partial-InternalFluxResponse**.

In a mathematical sense the MCA parameters are all derived by taking the single partial derivative of one variable or parameter with respect to another. Further insights into the control of a metabolic pathway can be obtained via additional coefficients that are obtained by second-order partial derivatives and the use of the chain rule of differentiation. All of the system parameters apply under the condition of a steady state of the concentrations of the reactants. On the other hand, the two elasticities apply to single enzymes and are fundamentally based on the form or nature of the steady-state rate equation; in many circumstances this equation will be a Michaelis-Menten equation or a simple elaboration of it.

Before turning to the application of the MCA methods described herein we consider the important task of estimating parameters from experimental observations of complex metabolic systems in cells.

5.8 Exercises

5.8.1

Consider the reaction scheme given in the first question/answer of Section 5.2. If the supply pathway has a flux control coefficient of 0.75 what will the flux control coefficients of the demand pathway be? What does this imply about the relative magnitudes of $\varepsilon_M^{v_d}$ and $\varepsilon_M^{v_s}$?

5.8.2

Consider the reaction scheme given in the first question/answer of Section 5.2. If the supply pathway has a concentration control coefficient of 0.5 what will be the concentration control coefficient of the demand pathway?

5.8.3

Use the **NMatrix** function introduced in Chapter 4 to verify that the stoichiometry matrix given for the reaction scheme in Eqn [5.8] is correct.

5.8.4

Calculate and interpret the flux control matrix for the reaction scheme given in Eqn [5.8] .

5.8.5

For the reaction scheme given in Eqn [5.8], calculate the value of $C_{v_1}^{S_3}$ by numerical perturbation.

5.8.6

The ε-elasticity matrix calculated in the first worked example in Section 5.4 has two non-zero entries in the second column. Why? What is the interpretation of the fact that the elements are of opposite sign?

5.8.7

Calculate the flux response matrix for the system presented in the first question/answer in Section 5.5. Are there any parameters which will have no effect on pathway flux at all?

5.9 References

1. Kascer, H. and Burns, J.A. (1973) The control of flux. *Symp. Soc. Exp. Biol.* **27**, 65-104.
2. Heinrich, R. and Rapoport, T.A. (1974) A linear steady-state treatment of enzymatic chains. General properties, control and effector strength. *Eur. J. Biochem.* **42**, 89-95.
3. Fell, D.A. (1992) MCA: a survey of its theoretical and experimental development. *Biochem. J.* **286**, 313-330.
4. Cornish-Bowden, A. (1995) *Fundamentals of Enzyme Kinetics.* Portland Press, London.
5. Heinrich, R. and Schuster, S. (1996) *The Regulation of Cellular Systems.* Chapman & Hall, New York.
6. Fell, D.A. (1997) *Understanding the Control of Metabolism.* Portland Press, London.
7. Hofmeyer, J.-H.S. and Cornish-Bowden, A. (1991) Quantitative assessment of regulation in metabolic systems. *Eur. J. Biochem.* **200**, 223-236.

6 Parameter Estimation

6.1 Introduction

Modelling any metabolic system involves three main tasks: (1) defining the mathematical functions that make up the individual units of the model, in other words, the rate equations of all the (bio)chemical reactions in the metabolic scheme; (2) collecting experimental kinetic and binding data on the various units of the system; and (3) deciding on values for the adjustable parameters in the various mathematical functions so that the agreement between the predictions of the model and the data are maximized. This latter task is known as *parameter estimation* or *model fitting*, and is the subject of this chapter.

The approach to modelling cellular systems used in this book is to develop models of the individual biochemical and biophysical processes, such as enzymic reactions and ion transport processes, and to collect these together to describe the system as a whole. Models of real cellular systems can become extremely complicated when constructed in this way. For example, the model of the red blood cell described in Chapter 7 contains 60 state variables (metabolites) and 270 parameters. Such systems are often said to be overparameterized with respect to the available experimental data. This is because in order to be able to estimate all the parameters in the model with a high degree of certainty, we need to follow all the independent state variables in the system during a very large number of experiments that involve system perturbations. Current experimental techniques do not enable this for most cellular systems. In other words, the rich parameterization of these models, combined with only sparse experimental data, does not allow reliable estimation of all the relevant parameter values.

On the other hand, when fitting a cellular model to experimental data, we usually have preconceptions of the values of many of the parameters involved. In the process of developing models of the individual enzymes and transporters, estimates are made of many of these parameters. Thus the problem of parameter estimation for the whole system is usually not one of *ab initio* estimation, but rather is one of refining prior estimates of parameter values. This approach to parameter estimation is discussed next. Note that the topic is very mathematical, and yet the various procedures that are discussed are implemented in *Mathematica* and are transparent to the user; hence this section can safely be skipped by those anxious to proceed to the modelling sections.

6.2 Approaches to Parameter Estimation

The aim of parameter estimation is to find the vector of values, φ, that leads to the best simulation of the given data, d. We seek a φ that has the maximum probability given the data, d; this is called the conditional probability, $p(\varphi|d)$. Bayes' theorem shows that this conditional probability can be factored into three terms :[1,2]

$$\max_{\varphi} \; p(\varphi \mid d) = \max_{\varphi} \; \frac{p(d \mid \varphi) \, p(\varphi)}{p(d)} \quad , \tag{6.1}$$

$$= \max_{\varphi} \; \frac{p(d \mid \varphi) \, p(\varphi)}{\int_{\varphi} p(d \mid \varphi) \, p(\varphi) \, d\varphi} \quad , \tag{6.2}$$

where $p(\varphi)$ is called the *prior probability* of φ, $p(d|\varphi)$ is the conditional probability of d given φ, and $p(d)$ describes the probability of the data. Since the denominator is an integral over all possible parameter values, it is a constant that is independent of the final φ. Hence,

$$\max_{\varphi} \; p(\varphi \mid d) = \max_{\varphi} \; p(d \mid \varphi) \, p(\varphi) \quad . \tag{6.3}$$

In other words, the particular φ that maximizes $p(\varphi|d)$ is found by maximizing the product of $p(d|\varphi)$ and $p(\varphi)$. Parameter estimation based on Eqn [6.3] is termed *maximum a posteriori* (MAP) estimation, and it plays a very important role in parameter estimation when modelling metabolic systems.

If we assume that $p(\varphi) = 1$, an assumption that we investigate in Section 6.4, then Eqn [6.3] reduces to

$$\max_{\varphi} \; p(\varphi \mid d) = \max_{\varphi} \; p(d \mid \varphi) \quad . \tag{6.4}$$

Parameter estimation using Eqn [6.4] is termed *maximum likelihood estimation* and it works by choosing parameter values that maximize the probability of the data. The least squares method of parameter estimation uses a maximum likelihood estimator, and this is discussed next.

6.3 Least Squares

Consider a data set d of N reaction velocities, v_i, each of which is associated with a particular substrate concentration, s_i; we believe that the function $v_i = v(s_i, \varphi)$ is an appropriate model to describe the data. Then the aim of parameter estimation (using Eqn [6.4]) is to maximize the probability of obtaining the N velocity measurements, given φ. In other words, we wish to estimate $\max_{\varphi} p(\varphi \mid d)$. To do this we must assign mathematical expressions that describe the errors in the measured data. It is usual to assume that each measurement is associated with an error that has a normal, or Gaussian, distribution around the true value $v(s_i, \varphi)$, with variance σ_i^2. Then the probability of observing a particular velocity v_i is given by

$$p(v_i) \, dv = \frac{1}{\sqrt{2\pi\sigma_i^2}} \, \mathrm{Exp}\left[\frac{-(v_i - v(s_i, \varphi))^2}{2\sigma_i^2} \right] dv \quad . \tag{6.5}$$

Assuming that every measurement is independent of every other one, the probability of the whole data set having the given values is the product of the probabilities of each data element, namely,

$$p(d \mid \varphi) = \prod_{i=1}^{N} \frac{1}{\sqrt{2 \pi \sigma_i^2}} \, \mathrm{Exp}\left[\frac{-(v_i - v(s_i, \varphi))^2}{2 \sigma_i^2} \right] dv \quad . \tag{6.6}$$

Now choose φ to maximize $p(d \mid \varphi)$. This is equivalent to maximizing the logarithm of the expression on the right-hand side of Eqn [6.6] or, more usefully, minimizing the negative of its logarithm:

$$\min_{\varphi} -\log p(d \mid \varphi) = \min_{\varphi} \sum_{i=1}^{N} \frac{(v_i - v(s_i, \varphi))^2}{2 \sigma_i^2} + \frac{1}{2} \, \log(2 \pi \sigma_i^2) \quad . \tag{6.7}$$

In Eqn [6.7] we can factor out the $1/2$ in the first term and we ignore the second term because it is independent of φ. Therefore our goal is to find values for the parameters that minimize

$$\chi^2 \equiv \sum_{i=1}^{N} \frac{(v_i - v(s_i, \varphi))^2}{\sigma_i^2} \quad . \tag{6.8}$$

This quantity is the "Chi-square" measure that is so familiar to users of statistics. It is important to remember that it is a maximum likelihood estimator that applies only to data with normally distributed errors.

6.4 Maximum *a Posteriori* (MAP)

In Section 6.2 we considered the concept of MAP estimation and the fact that it amounts to evaluating the expression

$$\max_{\varphi} p(\varphi \mid d) = \max_{\varphi} p(d \mid \varphi) \, p(\varphi) \quad . \tag{6.9}$$

In Section 6.3 we derived expressions for $p(d \mid \varphi)$ when the data have normally distributed errors. In order to use MAP we also need to be able to derive mathematical expressions for $p(\varphi)$. To do this we need to express, in mathematical terms, our prior knowledge about φ. With $\varphi = \{\varphi_1, \ldots, \varphi_M\}$, our prior knowledge will include knowledge about the distribution of each parameter. Typically we will have prior knowledge that each parameter, i, is normally distributed with a mean of $\overline{\varphi}_i$ and variance θ_i^2. Hence the probability that the value of parameter i is φ_i is given by

$$p(\varphi_i) \, d\varphi = \frac{1}{\sqrt{2 \pi \theta_i^2}} \, \mathrm{Exp}\left[\frac{-(\varphi_i - \overline{\varphi}_i)^2}{2 \theta_i^2} \right] d\varphi \quad . \tag{6.10}$$

If each parameter is statistically independent, then

$$p(\varphi) = \prod_{i=1}^{N} \frac{1}{\sqrt{2\pi\theta_i^2}} \mathrm{Exp}\left[\frac{-(\varphi_i - \overline{\varphi}_i)^2}{2\theta_i^2}\right] d\varphi \quad . \qquad [6.11]$$

If there are no prior expectations about the mean value of a parameter, a reasonable prior belief would be that the mean value is that which is estimated via MAP. Thus, $\overline{\varphi}_i$ becomes φ_i in Eqn [6.11] and hence $p(\varphi) = 1$. This step gives some justification to the use of the maximum likelihood estimator to derive the "chi-square" measure. Furthermore, if we have prior information only on the means of a subset of parameters, $\phi = \{\varphi_1, ..., \varphi_L\}$, such that $0 \le L \le M$, then

$$p(\varphi) = \prod_{i=1}^{L} \frac{1}{\sqrt{2\pi\theta_i^2}} \mathrm{Exp}\left[\frac{-(\varphi_i - \overline{\varphi}_i)^2}{2\theta_i^2}\right] d\varphi \quad . \qquad [6.12]$$

By combining Eqns [6.6] and [6.11] we obtain

$$p(\varphi \mid d) = \prod_{i=1}^{N} \frac{1}{\sqrt{2\pi\sigma_i^2}} \mathrm{Exp}\left[\frac{-(v_i - v(s_i, \varphi))^2}{2\sigma_i^2}\right] dv$$

$$\prod_{i=1}^{L} \frac{1}{\sqrt{2\pi\theta_i^2}} \mathrm{Exp}\left[\frac{-(\varphi_i - \overline{\varphi}_i)^2}{2\theta_i^2}\right] d\varphi \quad . \qquad [6.13]$$

As in Section 6.3, we maximize $p(\varphi|d)$ by minimizing the negative logarithm of Eqn [6.13]. Then by neglecting the terms that are independent of φ and by factoring out the various constants we obtain the result that $p(\varphi|d)$ can be maximized by minimizing the following expression:

$$\mathrm{MAP} \equiv \sum_{i=1}^{N} \frac{(v_i - v(s_i, \varphi))^2}{\sigma_i^2} + \sum_{i=1}^{L} \frac{(\varphi_i - \overline{\varphi}_i)^2}{\theta_i^2} \quad . \qquad [6.14]$$

In conclusion, we have shown (Sections 6.3 and 6.4) how when minimized two different functions yield parameter estimates that are the most probable for a particular data set. These functions are termed figure-of-merit functions (usually shortened to 'merit functions'). The first of these merit functions, χ^2, is appropriate for use when the prior information on the parameter estimates is minimal relative to the information provided by the data. This will be the case when estimating parameters for models of enzymic reactions and transport processes. The second of these functions, MAP, is useful when there is a significant amount of prior information on the probable values of the parameter set. This will be the case when constructing models of cellular systems based on models of the components of the system. We now turn to the techniques and algorithms that can be used to minimize these two functions.

6.5 Parameters in Rate Equations

Parameter estimation can be divided into two sections. The first involves estimating parameters in rate equations that describe enzymic reactions and transport processes. The second involves estimating parameters in models that consist of systems of differential equations. The methods chosen for the minimization of the merit function are different in each case because parameter estimation for a single rate equation usually employs techniques that rely on the determination of the gradient and second derivatives with respect to the parameters in the rate equation. On the other hand, for most systems of differential equations there is no analytical solution. In such cases analytical expressions for the first or second derivatives with respect to parameters are not available so we must rely on techniques that either do not use this information or that use numerical estimates of them.

6.5.1 Linear least squares

Most rate equations that describe biological processes are nonlinear. However, there are a few cases, such as passive transport of a solute across a membrane, that are describable as first-order processes. In addition, some nonlinear rate equations are readily transformed into linear ones; e.g., in Section 2.1 the linear Lineweaver-Burk plot was derived from the nonlinear Michaelis-Menten equation.

The *Mathematica* package, **Statistics`LinearRegression`**, contains several functions that perform linear least-squares estimation of parameters; the functions also deliver a number of associated statistical diagnostics. The function **Regress** performs a linear fit of an equation onto data by minimizing χ^2 and its default settings assume that the error distributions of each measurement have identical variances. However, the option **Weights** allows individual data variances to be included. The *Mathematica* Help browser gives further information on this package and its functions.

This package relies on matrix algebra and singular value decomposition and can yield the best fit parameters in a single step. The interested reader can find the theory behind these algorithms in a variety of sources.[2,3]

Let us now illustrate the process of linear least-squares estimation.

`<< `	load in the add −
`Statistics`LinearRegres`. `sion`	on package for performing linear regression
`Regress[`*data*`, {1, x}, {x}]`	Fits a linear model to *data*

Performing linear least squares regression.

Q: Show that linear regression can be used to estimate the parameters of the Michaelis-Menten rate equation if the rate data are first transformed according to the method of Lineweaver-Burk (Section 2.1.4).

A: We begin by loading the relevant *Mathematica* add-on package.

```
<< Statistics`LinearRegression`
```

Let us synthesize some "experimental data" that although not real are nonetheless realistic. They are generated by adding Gaussian noise to velocity data values derived from the Michaelis-Menten equation; we choose quite arbitrarily $V_{max} = 10$ mmol L^{-1} s^{-1} and $K_m = 3$ mmol L^{-1}. First, we generate the data without noise.

```
v₀[a_] := Vmax a / (Km + a) ;

Vmax = 10;
Km = 3;
dataEnz = Table[{a, v₀[a]}, {a, 0.5, 5, 0.5}]
```

```
{{0.5, 1.42857}, {1., 2.5}, {1.5, 3.33333},
 {2., 4.}, {2.5, 4.54545}, {3., 5.}, {3.5, 5.38462},
 {4., 5.71429}, {4.5, 6.}, {5., 6.25}}
```

Next we add Gaussian noise: to do this we open the add-on package **Statistics`ContinuousDistributions`** and then use the **Map, Random,** and **NormalDistribution** commands. The noise is assumed to have a mean of 0 and a standard deviation of 0.15.

```
<< Statistics`ContinuousDistributions`;
dataEnzNoisy = Map[
  {#[[1]], #[[2]] + Random[NormalDistribution[0, 0.15]]} &, dataEnz]
```

```
{{0.5, 1.45987}, {1., 2.39286}, {1.5, 3.54621},
 {2., 4.01127}, {2.5, 4.54339}, {3., 5.0744}, {3.5, 5.24343},
 {4., 5.66551}, {4.5, 6.04067}, {5., 6.08823}}
```

A "double-reciprocal" transformation is performed on **dataEnzNoisy** to implement Lineweaver-Burk analysis (Section 2.1.4) before **Regress** is used to fit a straight line to these transformed data.

```
dataDr = 1 / dataEnzNoisy;
fitR1 = Regress[dataDr, {1, x}, {x}]
```

""	"Estimate"	"SE"	"TStat"
$\{$ParameterTable $\to$ 1	0.10285`	0.004746`	21.6706`,
x	0.29359`	0.006027`	48.7059`

RSquared $\to$ 0.9966390404147704`,

AdjustedRSquared $\to$ 0.9962189204666166`,

EstimatedVariance $\to$ 0.00010056165576915086`$\}$

The default output of **Regress** is a table that describes various aspects of the regression process. The meanings of each of these components is given in the *Mathematica* Help browser or by consulting a standard statistics textbook (e.g., Rice[4]).

The fit yields estimates of the parameters that have small relative standard errors (SE), indicating a good fit. This can be verified further by plotting the fitted line over the data as follows. First, extract the parameters of the linear model from the **ParameterTable** in the output from **Regress**.

```
parTable = ParameterTable /. fitR1;
const = parTable[[1,1,1]];
slope = parTable[[1,2,1]];
```

Next, plot the data and model predictions on the same graph as follows:

```
graph1 = ListPlot[dataDr,
    PlotStyle -> {PointSize[0.02]}, DisplayFunction → Identity];
graph2 = Plot[const + x * slope, {x, -0.333, 2},
    AxesOrigin -> {0, 0}, DisplayFunction → Identity];
Show[{graph1, graph2}, DisplayFunction → $DisplayFunction,
    AxesLabel -> {"1/[a]", "1/v₀"}];
```

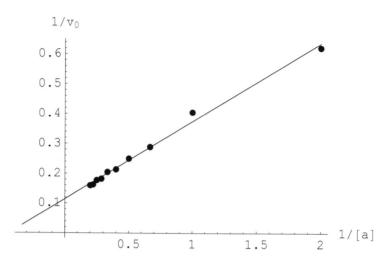

Figure 6.1. Lineweaver-Burk plot of synthetic noisy enzyme kinetic data. The ordinate is $1/v_0$ and the abscissa $1/[S]_0$.

Notice that the fitted line conforms well to the data.

Data transformations such as are used in the Lineweaver-Burk analysis also transform the error structure of the data. Consequently, they yield biased estimates of the parameters and are therefore only to be used for 'eye-balling' the data to assess conformity of the model with the data, and for providing initial estimates of parameter values for nonlinear regression analysis. This method which is described next should be used for obtaining the definitive estimates of parameter values and their associated errors.

6.5.2 Nonlinear estimation

As stated in the previous section, the majority of rate equations that describe cellular systems are nonlinear with respect to their dependent variables and their parameters. Unfortunately this means that we can no longer use the techniques of linear algebra and so we rely on iterative-search methods for parameter estimation. For nonlinear estimation we start with an initial trial set of parameter values and progressively refine these by iteratively applying an algorithm that adjusts the values so as to decrease the value of a merit function.

The Levenberg-Marquardt algorithm is the standard one used in nonlinear parameter estimation. It relies on two different methods for the minimization of the merit function. The first is the steepest descent, or gradient descent, method which decreases the value of the merit function by using information on the direction in which the rate of decrease is the greatest. For example, if χ^2 (Eqn [6.8]) is the merit function, then the direction of steepest descent is given by the negative gradient, $-\nabla \chi^2 (\varphi)$, where

$$(\nabla \chi^2)_j = \frac{\partial \chi^2}{\partial \varphi_j} = -2 \sum_{i=1}^{N} \frac{v_i - v(s_i, \varphi)}{\sigma_i^2} \frac{\partial v(s_i, \varphi)}{\partial \varphi_j} \ . \qquad [6.15]$$

By making a step α in the direction $-\nabla \chi^2 (\varphi)$, φ_{next} is estimated from $\varphi_{currnet}$ using

$$\varphi_{next} = \varphi_{current} - \alpha \nabla \chi^2 (\varphi_{current}) \ . \qquad [6.16]$$

The second method of determining φ_{next} from $\varphi_{current}$ is to evaluate the merit function using a second-order Taylor approximation around $\varphi_{current}$. Again, taking Eqn [6.8] as the merit function, the second-order approximation of χ^2 around $\varphi_{current}$ is

$$\chi^2 (\varphi) = \chi^2 (\varphi_{current}) + $$
$$\nabla \chi^2 (\varphi_{current}).(\varphi - \varphi_{current}) + \frac{1}{2} (\varphi - \varphi_{current}).\mathbf{H}.(\varphi - \varphi_{current}) \ , \qquad [6.17]$$

where **H** is the matrix of second derivatives called the Hessian. It is defined as follows:

$$\mathbf{H}_{jk} = \frac{\partial^2 \chi^2 (\varphi)}{\partial \varphi_j \partial \varphi_k} = $$
$$2 \sum_{i=1}^{N} \frac{1}{\sigma_i^2} \left[\frac{\partial v(s_i, \varphi)}{\partial \varphi_j} \frac{\partial v(s_i, \varphi)}{\partial \varphi_k} - (v_i - v(s_i, \varphi)) \frac{(\partial)^2 v(s_i, \varphi)}{\partial \varphi_j \partial \varphi_k} \right] \ . \qquad [6.18]$$

In this case φ_{next} is selected to be the set of parameter values which minimizes Eqn [6.17]. This set is found by setting $\nabla \chi^2 (\varphi) = 0$; in other words, we solve

$$\nabla \chi^2 (\varphi_{next}) = \chi^2 (\varphi_{current}) + \mathbf{H}.(\varphi_{next} - \varphi_{current}) = 0 \qquad [6.19]$$

for φ_{next}. The solution is

$$\varphi_{next} = \varphi_{current} - \mathbf{H}^{-1}.\nabla \chi^2 (\varphi_{next}) \ . \qquad [6.20]$$

The method is called Newton's method or the inverse-Hessian method.

The reason that the Levenberg-Marquardt method employs the two iterative methods is that the reliability and efficiency of each method vary depending on how close the initial parameter estimates are to their final values. For example, if the initial estimate of $\varphi_{current}$ is very close to the global minimum the inverse-Hessian method converges quickly; while the steepest descent method converges relatively slowly because the gradient becomes progressively smaller as the minimum is approached. Alternatively, if the estimate of $\varphi_{current}$ is far from the minimum, the value given for φ_{next} by the inverse-Hessian method can be even farther away from the minimum than the previous $\varphi_{current}$.

How does the Levenberg-Marquardt algorithm 'know' when to switch between the two methods? The answer is as follows: the algorithm first uses the diagonal elements of the

Hessian matrix to specify the order of magnitude of the step (i.e., α in Eqn [6.16]) that is taken in the steepest descent method, hence Eqn [6.16] becomes

$$\varphi_{i,\text{next}} = \varphi_{i,\text{current}} - \frac{1}{\lambda H_{ii}} \frac{\partial \chi^2}{\partial \varphi_i} \quad, \tag{6.21}$$

where λ is a scalar factor. The choice of α is based on the curvature of the error surface given by H_{ii}. If the curvature is large then very small gradient descent steps are taken.

Having defined the size of the gradient descent steps, α, the algorithm combines the steepest descent and inverse-Hessian methods by defining a new matrix, $\mathbf{M}$, such that

$$M_{ii} = \frac{1}{2} H_{ii} (1 + \lambda) \quad \text{and} \quad M_{ij} = \frac{1}{2} H_{ij} \quad j \neq i \tag{6.22}$$

This allows the replacement of Eqns [6.16] and [6.20] with

$$\varphi_{\text{next}} = \varphi_{\text{current}} - \mathbf{M}^{-1}.\nabla \chi^2 (\varphi_{\text{next}}) \tag{6.23}$$

so that when λ approaches zero Eqn [6.23] approaches the inverse-Hessian method, and when λ is relatively large, $\mathbf{M}$ becomes diagonally dominant and Eqn [6.23] approaches the steepest descent method.

In conclusion, the usual approach to implementing the Levenberg-Marquardt algorithm is to begin with a moderate value for λ and solve Eqn [5.20]. Then if the value of the merit function decreases in size, λ is decreased (say, by a factor of 10) forcing Eqn [5.20] toward the inverse-Hessian method. On the other hand, if the merit function increases in value, then λ is increased by a factor of 10 forcing Eqn [5.20] toward the steepest descent method.

6.5.3 Nonlinear least squares

The *Mathematica* add-on package **Statistics`NonlinearFit** contains several nonlinear least squares functions which use the Levenberg-Marquardt method as the default option. One of these functions, **NonlinearRegress**, is illustrated next.

<<	load in the add-on package for
Statistics`NonlinearFit	performing nonlinear least-squares estimation
NonlinearRegress[*data*, {1, *x*}, {*x*}]	Fits a linear model to *data*

Performing nonlinear least-squares regression analysis.

Q: How do we use nonlinear regression analysis to fit the Michaelis-Menten function onto enzyme kinetic data?

A: We illustrate the method by using the kinetic data from the previous example; it is labelled **dataEnzNoisy**.

First, load the **NonlinearFit** package.

```
<< Statistics`NonlinearFit`;
Clear[Subscript]
```

Next, use the **NonlinearRegress** function as follows.

$$\text{results} = \textbf{NonlinearRegress}\left[\textbf{dataEnzNoisy}, \frac{V_{max}\ a}{K_m + a}, a, \{V_{max}, K_m\}\right]$$

$\big\{$BestFitParameters → $\{V_{max}$ → 9.66041, K_m → 2.7907$\}$,

ParameterCITable →

	Estimate	Asymptotic SE	CI
V_{max}	9.66041	0.274669	{9.02702, 10.2938},
K_m	2.7907	0.168322	{2.40255, 3.17885}

EstimatedVariance → 0.00829601, ANOVATable →

	DF	SumOfSq	MeanSq
Model	2	217.467	108.733
Error	8	0.0663681	0.00829601,
Uncorrected Total	10	217.533	
Corrected Total	9	21.9156	

$$\text{AsymptoticCorrelationMatrix} \rightarrow \begin{pmatrix} 1. & 0.97612 \\ 0.97612 & 1. \end{pmatrix},$$

FitCurvatureTable →		Curvature
	Max Intrinsic	0.0235299
	Max Parameter-Effects	0.106493
	95. % Confidence Region	0.473568

The meaning of the various elements of this *Mathematica* output are found in the Help Browser and in standard statistical textbooks, so we will not discuss them in detail here. However, it is worth noting that the key numbers are the parameter estimates and the "Asymptotic SE" values; these are the standard errors of the parameter estimates which provide a quantitative idea of their reliability.

The synthetic data and the estimated Michaelis-Menten equation can be plotted together, as follows:

```
graph1 = ListPlot[dataEnzNoisy,
    PlotStyle -> {PointSize[0.02]}, DisplayFunction → Identity];
graph2 = Plot[ Vmax a/(Km + a) /. {BestFitParameters /. results},
    {a, 0, 10}, DisplayFunction → Identity];
Show[{graph1, graph2}, DisplayFunction → $DisplayFunction,
    AxesLabel -> {"Time (h)", "v0"}];
```

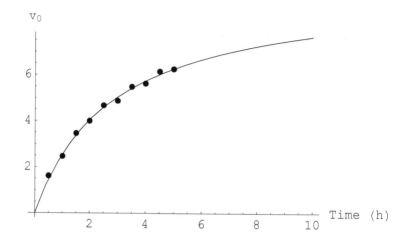

Figure 6.2. Nonlinear regression of the Michaelis-Menten expression onto enzyme kinetic data using the *Mathematica* function **NonlinearRegress**. The ordinate is v_0 and the abscissa $[S]_0$.

6.5.4 Nonlinear MAP

The algorithms supplied in the package **Statistics`NonlinearFit`** only perform parameter estimation based on least-squares minimization. On the other hand, the function **FindMinimum** uses a variety of methods, including the Levenberg-Marquardt method, to determine the local minima of a user defined function. So, if we define our own MAP function we can use **FindMinimum** for parameter estimation, as follows:

FindMinimum[	Searches for a local minimum
$f, \{x, x_0\}, \{y, y_0\}$]	in a function of several variables

Minimizing merit functions.

Q: Is it possible to fit the Michaelis-Menten equation onto enzyme kinetic data assuming that we have prior knowledge about the V_{max} of the enzyme? Suppose that the prior knowledge is that the errors in V_{max} are normally distributed with a mean of 10 mmol L^{-1} min^{-1} and a variance that is 10 times less than the variance of the experimental data. The synthetic data from the previous example (**dataEnzNoisy**) fulfills these criteria, since they were constructed with these properties in the first place.

A: We have prior knowledge about the value of V_{max} so we use MAP estimation rather than least-squares estimation. First, define the MAP merit function, recalling that a MAP function for normally distributed measurement errors and normally distributed prior parameter probabilities is given by Eqn [6.14].

Hence for the data in **dataEnzNoisy** the MAP merit function is given by Eqn [6.14], as follows:

```
v₀[a_] := Vmax a ;
         ───────
         Km + a

conc = Transpose[dataEnzNoisy][[1]];
velMeas = Transpose[dataEnzNoisy][[2]];
velPred = Map[v₀, conc];
MAP[Km_, Vmax_] :=
   Plus @@ (velMeas - velPred)² + 10 * (Vmax - 10)²;
```

Now apply the **FindMinimum** function to estimate the Michaelis-Menten parameters.

```
FindMinimum[MAP[Km, Vmax], {Vmax, 10.}, {Km, 3.}]

{0.0778725, {Vmax → 9.99674, Km → 2.99311}}
```

Notice how the estimate of V_{max} is much closer to 10 than in the previous example. The fact that the variance of our prior distribution for V_{max} was low compared to the experimental data means that deviations from this prior belief were heavily weighted in the MAP merit function.

The important topic of estimating the standard errors associated with parameter estimates that are obtained by using MAP are discussed in Section 6.8.

6.6 Parameters in Systems of Differential Equations

Kinetic models of cellular systems are usually described by arrays of simultaneous differential equations. There are two features of these arrays that make parameter estimation very difficult. First, the number of parameters in the models tends to be large which means that the process of optimization of parameters can be very slow. More will be discussed about this aspect in Section 6.7. Second, for most realistic models there is no analytical solution to the array of differential equations. Thus, unlike the situation with single rate equations it is not possible to determine analytical expressions for the first or second derivatives of the rate equations with respect to the various parameters in the array. Therefore we must rely on techniques that either do not require first or second derivatives (direct methods) or which use numerical estimates of these values. Because of the high time–cost involved in numerically estimating derivatives, direct methods are usually preferred; these are described next.

The two main approaches to minimizing merit functions without evaluating derivatives are (1) the downhill simplex method and (2) the 'direction-set' modification of Powell's method. *Mathematica*'s **FindMinimum** function relies on a modification of Powell's method so is the one explored here. For more details on the downhill simplex method, see Press et al.[3]

The basic idea behind the direction-set method is to first minimize the objective function by changing the values of the parameter estimates in the direction of a chosen vector. Once the minimum along this vector direction (line) has been found, a new vector direction is determined and the merit function minimized in this new direction. By a series of such line minimizations, the algorithm iteratively locates the minimum of the merit function. The 'trick' to successful direction-set methods lies in the way they calculate the directions for each successive line minimization; Press et al.[3] contains the mathematical details.

In *Mathematica* **FindMinimum** automatically uses a direction-set method if derivatives cannot be found. The use of **FindMinimum** is illustrated next.

Q: Suppose we have noisy experimental data consisting of reactant concentrations over a 10-h time course from a linear reaction scheme like that in Eqn [6.24]. Also, suppose that prior information on the error estimates of these concentrations indicates that the noise is normally distributed with a mean of zero and a standard deviation of 0.03 units. How can we use the data to obtain a best estimate of the kinetic parameter values ($K_{m,i}$ and $V_{max,i}$, $i = 1,...,3$) in the model?

$$S_1 \xrightarrow{v_1} S_2 \xrightarrow{v_2} S_3 \xrightarrow{v_3} S_4 \qquad\qquad [6.24]$$

Aside: For this exercise we must first synthesize the noisy data. To do this we make each v_i a simple irreversible Michaelis-Menten rate equation (Section 2.3.1) with all V_{max} values being 1 mmol L^{-1} h^{-1}, all K_m values 1 mmol L^{-1}, and $S_1[0] = 1$ mmol L^{-1}, $S_2[0] = S_3[0] = S_4[0] = 0$ mmol L^{-1}. For the subsequent analysis our prior knowledge of the mean of the parameters will be the parameter values used to calculate the synthetic data in the first place; and recall that the synthetic data are made noisy with normally distributed random variations that have a mean of 0 and standard deviation of 0.03.

A: MAP estimation is used for this problem. The first step entails defining the equations of the model in the form of a mathematical function that is amenable to processing via **FindMinimum**. Hence, the model must be set up as a system of matrices, as is done in Chapter 4. To start this analysis we reload the package **MetabolicControl-**

Analysis and specify the various initial substrate concentrations, stoichiometry matrix, and rate equations.

```
<< MetabolicControlAnalysis`
```

```
S̄ := Table[sᵢ[t], {i, 4}];
```

$$\bar{N} = \begin{pmatrix} -1 & 0 & 0 \\ 1 & -1 & 0 \\ 0 & 1 & -1 \\ 0 & 0 & 1 \end{pmatrix};$$

$$v[1] := \frac{Vm1\ s_1[t]}{Km1\ (1 + \frac{s_1[t]}{Km1})};$$

$$v[2] := \frac{Vm2\ s_2[t]}{Km2\ (1 + \frac{s_2[t]}{Km2})};$$

$$v[3] := \frac{Vm3\ s_3[t]}{Km3\ (1 + \frac{s_3[t]}{Km3})};$$

```
v̄ := Table[v[i], {i, 3}];

ic1 = {s₁[0] == 1, s₂[0] == 0, s₃[0] == 0, s₄[0] == 0};

pars = {Vm1, Vm2, Vm3, Km1, Km2, Km3};
parValues = {1, 1, 1, 1, 1, 1};
```

Detour: We interrupt our analysis to generate the "experimental" data that we will subsequently analyze. We define a function that produces a table of simulated substrate concentrations starting at time = 0 h, continuing in 1-h increments to 10 h, based on Eqn [6.24]. The argument of the function is a list of parameter values. To define the function we use the **Module** command noting that this function is not self-contained and requires values that are defined in the Cell above.

```
model[a_List] :=
 Module[{parTable, result, sol},
  parTable = Transpose[{pars, a}];
  sol = NDSolveMatrix[S̄, N̄, v̄,
    ic1, {t, 0, 10}, parTable, MaxSteps → 10000];
  result = Table[Evaluate[S̄ /. sol], {t, 0, 10, 1}]
 ]
```

Having defined this function, and after having loaded the **Statistics`ContinuousDistributions`** package, we can create the synthetic data.

```
<< Statistics`ContinuousDistributions`;
exptData = model[parValues];
noisyExptData =
  MapAll[# + Random[NormalDistribution[0, 0.03]] &, exptData];
plotData = Table[{{i - 1, noisyExptData[[i,1,1]]},
    {i - 1, noisyExptData[[i,1,2]]}, {i - 1, noisyExptData[[i,1,3]]},
    {i - 1, noisyExptData[[i,1,4]]}}, {i, 1, 10}];
ListPlot[Flatten[plotData, 1],
  AxesLabel -> {"Time (h)", "Concentration (mM)"},
  PlotStyle -> {PointSize[0.02]}];
```

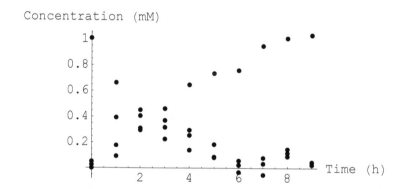

Figure 6.3. Plot of the synthetic experimental data.

Return from detour: Having obtained the experimental data we now analyze it to obtain estimates of the parameter values. Using the function in Eqn [6.25] we define the merit functions for the least squares and MAP estimations, as follows:

```
ss[a_List] :=
  Plus @@ (Flatten[noisyExptData] - Flatten[model[a]]) ^ 2

map[a_List] :=
  Plus @@ (Flatten[noisyExptData] - Flatten[model[a]]) ^ 2 +
    Plus @@ (a - {1, 1, 1, 1, 1, 1}) ^ 2
```

The least-squares parameter estimates are found by minimizing the **ss[a]** (sum-of-squares) merit function; but when we use **FindMinimum** on a function that does not have derivatives it is necessary to specify an upper and lower bound of the initial values for each parameter.

```
ssResults = FindMinimum[ss[pars], {Vm1, 1, 2},
  {Vm2, 1, 2}, {Vm3, 1, 2}, {Km1, 1, 2}, {Km2, 1, 2},
  {Km3, 0.1, 10}, PrecisionGoal → 3, MaxIterations → 100]

{0.946883, {Vm1 → -18.8473, Vm2 → 1.03136,
  Vm3 → 1., Km1 → 1., Km2 → 1., Km3 → -0.091883}}
```

Result: It is evident in this simple example that the synthetic data do not allow the accurate estimation of all the parameter values. However, if we use MAP estimation by minimizing **map[a]**, as is shown next, the parameters that were used to create the synthetic data are recovered.

```
mapResults = FindMinimum[map[pars], {Vm1, 1, 2},
  {Vm2, 1, 2}, {Vm3, 1, 2}, {Km1, 1, 2}, {Km2, 1, 2},
  {Km3, 1, 2}, PrecisionGoal → 3, MaxIterations → 100]

{0.946883,
  {Vm1 → 1., Vm2 → 1., Vm3 → 1., Km1 → 1., Km2 → 1., Km3 → 1.}}
```

This is a remarkable result which could be dwelt upon much more via further worked examples, but restrictions on space do not allow this. However, we hope that you will be able to generate your own examples using the above Cells as a template for the analytical procedures.

6.7 Optimal Parameters

A problem with the iterative methods of nonlinear parameter estimation is that they are only useful for finding local minima. They will only locate a minimum in a merit function that is near that defined by the starting values of the parameters. By performing parameter estimation from a single starting set we have no way of knowing whether the minimum in the merit function that has been found is a global one. Therefore it is important to perform the merit function minimization by using a number of randomly generated starting parameter sets. If the same minimum is arrived at from each starting set we often assume that the true global minimum has been located. If different minima are located a number of useful approaches can be used, such as simulated annealing and genetic algorithms. These methods are beyond the intended scope of this book, but the interested reader can find the details in Gershenfeld.[2] The implementation of these analyses in *Mathematica* requires some complicated programming.

6.8 Variances of Parameters

Calculating the uncertainties in parameter estimates is a very important and yet very technical topic. Fortunately, the *Mathematica* functions **Regress** and **NonlinearRegress** automatically yield the standard errors of the estimates. On the other hand, when **FindMinimum** is used to minimize a merit function, no error estimates are given. So it is important to have a strategy to make these estimates. This can be done via a so-called Monte Carlo simulation.

The idea behind the Monte Carlo approach is to assume that the parameters that have been estimated are the true underlying parameters relevant to the real system. Then these parameter values, along with a function that generates noise, is used with the model to simulate sets of synthetic data. This is done in exactly the manner used in the example above. Several simulated data sets are then subjected to parameter estimation and from the set of estimates for each parameter a mean and standard deviation is calculated, using the conventional formulae.

Q: What are the standard deviations of the least-squares estimates of the parameters obtained for the "experimental" data from the metabolic model of the previous example?

A: The estimate of the standard deviations uses the Monte Carlo simulation technique. Recall that the results of the least-squares estimation are called **ssResults**.

```
ssResults
```

```
{0.946883, {Vm1 → -18.8473, Vm2 → 1.03136,
    Vm3 → 1., Km1 → 1., Km2 → 1., Km3 → -0.091883}}
```

Use the following input to simulate a single set of synthetic data using these results.

```
dataS1 = model[pars /. ssResults[[2]]];
dataS2 = Flatten[dataS1, 1];
synData =
  MapAll[# + Random[NormalDistribution[0, 0.03]] &, dataS2];
```

To simulate an array of 25 data sets we use

```
Do[
  syntheticData[i] = MapAll[
    # + Random[NormalDistribution[0, 0.03]] &, dataS2], {i, 25}]
```

The merit function for each **syntheticData[i]** is

```
ssNew[i_, a_List] :=
  Plus @@ (Flatten[syntheticData[i]] - Flatten[model[a]]) ^ 2
```

Next, apply **FindMinimum** to each of the 25 merit functions.

```
Do[
  ssRes[i] = FindMinimum[ssNew[i, pars], {Vm1, 1, 2}, {Vm2, 1, 2},
    {Vm3, 1, 2}, {Km1, 1, 2}, {Km2, 1, 2}, {Km3, 1, 2},
    PrecisionGoal → 2, MaxIterations → 1000], {i, 25}]
```

Finally, by using the package **Statistics`MultiDescriptiveStatistics`**, the standard deviation of each parameter estimate is determined.

```
<< Statistics`MultiDescriptiveStatistics`
data = Table[pars /. ssRes[i]⟦2⟧, {i, 25}];
sd = StandardDeviation[data];
{pars, pars /. ssResults⟦2⟧, sd} // MatrixForm
```

$$
\begin{pmatrix}
Vm1 & Vm2 & Vm3 & Km1 & Km2 & Km3 \\
-18.8473 & 1.03136 & 1. & 1. & 1. & -0.091883 \\
11.0746 & 10.7785 & 0.0195382 & 0.0329097 & 0.0767816 & 0.00239576
\end{pmatrix}
$$

This is a very neat outcome and essentially provides a recipe for estimating parameter values, and their standard deviations, for models that describe very complex systems of metabolic reactions. Such complexity is encountered in the next two chapters.

6.9 Exercises

6.9.1

Determine the steady-state parameters for the enzyme kinetic data given in the worked example in Section 6.5.1 using a Eadie-Hofstee plot and the linear regression package in *Mathematica*.

6.9.2

Re-run the example given in Section 6.5.1 with a larger standard deviation for the noise function. What happens to the relative size of the error estimates on the parameters?

6.9.3

The noise function used for the velocity data for the example given in Section 6.5 had a constant standard deviation. This is the situation for many experimental techniques such as NMR spectroscopy; with NMR the noise in a spectrum has a fixed value but the signal intensity (peak area) depends on the relative amounts of compounds present in the sample. Thus the signal-to-noise ratio of the spectral peaks varies from one peak to the next, depending on the concentration of the substances.

On the other hand, in some experimental data the error term has a standard deviation that is a constant proportion of the measured velocity. Reanalyze the example in Section 6.5.1 with such an error model.

6.9.4

For the example in Section 6.5.3, examine the output of the **NonlinearRegress** function to determine 95% confidence intervals for each parameter. Do you understand how these were calculated?

6.9.5

Re-run the worked example in Section 6.5.4 assuming that we only have prior information about the K_m value and not V_{max}. Assume that the prior distribution for K_m has a mean of 3 mmol L^{-1} and a variance five times less than the variance of the synthetic experimental data.

6.9.6

Use the Monte Carlo simulation method to determine the standard deviations of the estimated parameters in the worked example in Section 6.5.4.

6.10 References

1. Bernardo, J.M. and Smith, A.F.M. (1994) *Bayesian Theory*, Wiley, New York.
2. Gershenfeld, N. (1999) *The Nature of Mathematical Modeling*, Cambridge University Press, New York.
3. Press, W.H., Teukolsky, S.A., Vetterling, W.T., and Flannery, B.P. (1992) *Numerical Recipes in C: The Art of Scientific Computing*. 2nd ed., Cambridge University Press, New York.
4. Rice, J.A. (1995) *Mathematical Statistics and Data Analysis*. 2nd ed., Duxbury Press, Belmont, California.

7 Model of Erythrocyte Metabolism

7.1 Introduction

In the earlier chapters of this book we aimed to show how *Mathematica* can be used to simulate some relatively simple models of metabolism. We surveyed a lot of territory and presented a large number of analytical tools. To some readers, the analyses, especially in Chapter 5, may have seemed rather esoteric, with their heavy use of mathematics applied to simple and possibly unrealistic models. In this and the next chapter we aim to show that it was not mathematics for its own sake, but that the analytical procedures can give valuable insights that cannot be obtained in any other way.

In the following sections we tackle much more complicated real-life problems; namely, we present and analyze a realistic model of erythrocyte metabolism. The model is one that was developed for use in our day-to-day experimental work.[1–4] It has been used as an aid to experimental design and for the analysis of experimental data. In addition to providing an illustrative example of modelling complex systems it is hoped that the human erythrocyte model, and its associated enzyme kinetic equations, will be a useful catalogue of kinetic equations for those working in the area of metabolic biochemistry.

7.2 Models of Erythrocyte Metabolism

Over the last 25 years many mathematical models of erythrocyte metabolism have been developed.[4–7] These have been very successful in identifying the key features of the regulation and control of the metabolism of the cell. Indeed the erythrocyte is the best modelled of all biochemical systems and there are a number of reasons for this. The first is the ease of obtaining them by simple venipuncture. Second, the erythrocyte has relatively simple metabolism as a result of lacking mitochondria and other organelles (Fig. 7.1).

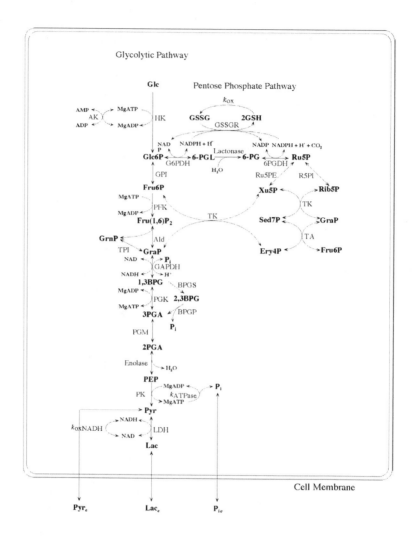

Figure 7.1. Erythrocyte metabolism. The best known physiological function of the erythrocyte is the facilitation of oxygen transport throughout the body. In response to this task, the erythrocyte has evolved into a highly specialized but metabolically simple cell. The mature human erythrocyte has lost all organelles and hence its metabolism is primarily reduced to the glycolytic and pentose phosphate pathways. These two pathways generate the ATP and reducing equivalents that keep the cell functionally active. A peculiar feature of glycolysis in erythrocytes is the possession of an alternative pathway for carbon flux via 1,3-bisphosphoglycerate (1,2-BPG). This pathway, known as the 2,3-BPG or Rapoport-Luebering shunt, bypasses phosphoglycerate kinase by converting 1,3-BPG to 2,3-BPG in the metabolic pathway. The enzymes are denoted by: AK (adenylate kinase); Ald (aldolase); BPGP (2,3-BPG phosphatase); BPGS (2,3-BPG synthase); G6PDH (glucose-6-phosphate dehydrogenase); GAPDH (glyceraldehyde-3-phosphate dehydrogenase); GPI (glucosephosphate isomerase); HK (Hexokinase); kATPase (non-glycolytic energy consumption); kox (reduction processes consuming GSH); koxNADH

(reducing processes requiring NADH); Lactonase (δ-gluconolactonase); LDH (lactate dehydrogenase; note that the model also includes an NADPH-dependent lactate dehydrogenase); PFK (phosphofructokinase); PGK (phosphoglycerate kinase); 6PGDH (6-phosphogluconate dehydrogenase); PGM (phosphoglycerate mutase); PK (pyruvate kinase); R5PI (ribose-5-phosophate isomerase); Ru5E (ribulose-5-phosphate epimerase); TA (transaldolase); TK (transketolase); TPI (triose phosphate isomerase). Metabolites: 1,3-BPG (1,3-bisphosphoglycerate); 2,3-BPG (2,3-bisphosphoglycerate); Ery4P (erythrose 4-phosphate); Fru(1,6)P2 (fructose 1,6-bisphosphate); Fru6P (fructose 6-phosphate); Glc (glucose); Glc6P (glucose 6-phosphate); GraP (glyceraldehyde 3-phosphate); GrnP (dihydroxyacetone phosphate); Lac (lactate); Lace (extracellular lactate); Pi (inorganic phosphate); Pie (extracellular inorganic phosphate); PEP (phosphenolpyruvate); 2-PGA (2-phosphoglycerate); 3-PGA (3-phosphoglycerate); 6-PG (6-phosphogluconate); 6-PGL (6-phosphoglucolactone); Pyr (pyruvate); Pyre (extracellular pyruvate); Rib5P (ribose 5-phosphate); Ru5P (ribulose 5-phoshate); Sed7P (sedoheptulose 7-phosphate); Xu5P (xylulose 5-phosphate).

The model presented in this chapter was developed with the primary aim of illuminating our understanding of the regulation and control of the 2,3-bisphosphoglycerate (2,3-BPG) concentration. 2,3-BPG is important in regulating blood oxygen transport and delivery. In clinical and environmental conditions where oxygen transport has been compromised, such as anemia, congenital heart disease, and high altitude, the concentration of 2,3-BPG is often elevated well above normal values. 2,3-BPG is a heterotropic allosteric effector of oxygen binding by hemoglobin (Hb). By binding preferentially to the deoxygenated form of Hb, it decreases the apparent affinity of Hb for O_2. Although the interactions between Hb and 2,3-BPG had been known for more than 30 years prior to the development of this model, the precise regulatory features of 2,3-BPG metabolism were still an issue of much debate until recently.

7.3 Stoichiometry of Human Erythrocyte Metabolism

The model is based on three main metabolic systems: glycolysis, the pentose phosphate pathway (PPP), and the 2,3-BPG shunt (Fig. 7.1). It also includes the transport of lactate, pyruvate, and phosphate across the plasma membrane. Glucose transport is also included but the rate of this process is known to be so rapid in humans that the concentration of intracellular glucose in the model is set equal to that of extracellular glucose. A number of processes that are particularly important in relation to 2,3-BPG metabolism are simulated and these include the binding of metabolites to hemoglobin and magnesium ions, as well as the effects of pH on several enzymic reactions and binding processes. A final point worth noting is that, in the first presentation of the model, the intracellular volume is assumed to be constant.

The rate equations for all these processes are presented in Appendix 3, 'Rate Equations for Enzymes of the Red Cell.' The equations can readily be up-loaded for use in the current *Mathematica* session. Note, however, that before these equations can be called the reader must evaluate all the cells in Appendix 3 so that the appropriate files are created.

In order to construct the model using these rate equations, we must first specify the stoichiometry matrix, and the substrate and reaction vectors. In addition, since this model involves two different compartment volumes, the intracellular volume ($\mathbf{Vol_i}$) and the extracellular volume ($\mathbf{Vol_e}$), we would normally have to specify a diagonal volume matrix, $\mathbf{V}$, as was done in Section 4.8. However, in the worked example (question/answer) below we generate the matrix ($\mathbf{V^{-1}\,N}$) (see Section 4.8) in one step. The little trick we use to do this may appear difficult to understand at first, but it greatly increases the speed of construction of the model.

Q: Specify the stoichiometry of the model of erythrocyte metabolism: this step will provide the specification of all of the rate equations and hence the interconnectedness of the metabolic reactions.

A: The first action is to load the red cell kinetic equations that are in **RBCequations** and the **MetabolicControlAnalysis** add-on package.

```
<< RBCequations;
<< MetabolicControlAnalysis`
```

The next step that would normally be used in model construction would be to specify the stoichiometry of the model by writing down the list of all the reactions that will form part of the model. We would then use **NMatrix** to define $\mathbf{N}$, and then separately define the diagonal volume matrix, $\mathbf{V}$, associated with $\mathbf{N}$. We can, however, define the matrix ($\mathbf{V^{-1}\,N}$) in one step by using the following trick. This involves defining the list of reactions as usual, but in addition, multiplying the stoichiometry coefficient of each metabolite by the volume of the compartment in which the metabolite is found.

The trick is only required for reactions that explicitly or implicitly involve metabolite transport. Thus for these reactions we will express the reaction rates in terms of the rate of change of *amounts*, while all other reaction rates will be expressed in terms of the rate of change of *concentration*. The rate equations that we load with the command **<<RBCequations** were defined following these conventions.

To use this method the list of reactions which make up the model is as follows. Note that each reaction is given a name that corresponds to its name in the **RBCequations** file.

```
eqns = {
   (*Glycolytic reactions.*)
   {hk,    Glc[t]          +  MgATP[t]    →  Glc6P[t]   +   MgADP[t]   },
           ─────────────      ────────       ────────       ────────
           Voli + Vole         Voli           Voli           Voli
   {gpi,   Glc6P[t]   →   Fru6P[t]} ,
   {pfk,   Fru6P[t] + MgATP[t]   →   Fru16P2[t] + MgADP[t]},
   {ald,   Fru16P2[t]   →   GrnP[t]   + GraP[t]},
   {tpi,   GraP[t]  →  GrnP[t] },
   {gapdh, GraP[t] + Phos[t] + NAD[t]  → B13PG[t] + NADH[t]},
```

{pgk,   B13PG[t] + MgADP[t] → P3GA[t] + MgATP[t]},

{pgm,   P3GA[t] → P2GA[t]},

{eno,   P2GA[t] → PEP[t]},

{pk,   PEP[t] + MgADP[t] → Pyr[t] + MgATP[t]},

{ldh,   Pyr[t] + NADH[t] → Lac[t] + NAD[t] },

{ldhp,   Pyr[t] + NADPH[t] → Lac[t] + NADP[t]},

(*Reactions of 2,3 BPG synthase-phosphatase.*)

{bpgsp1,   B13PG[t] + BPGSP[t] ↔ BPGSP$B13PG[t] },

{bpgsp2,   BPGSP$B13PG[t] → BPGSPP[t] + P3GA[t]},

{bpgsp3,   BPGSPP[t] + P3GA[t] → BPGSPP$P3GA[t] },

{bpgsp4,   BPGSPP[t] + P2GA[t] → BPGSPP$P2GA[t]},

{bpgsp5,   BPGSPP$P3GA[t] → BPGSP$B23PG[t] },

{bpgsp6,   BPGSPP$P2GA[t] → BPGSP$B23PG[t] },

{bpgsp7,   BPGSP$B23PG[t] → BPGSP[t] + B23PG[t] },

{bpgsp8,   BPGSPP[t] + Phos[t] → BPGSPP$Phos[t] },

{bpgsp9,   BPGSPP$Phos[t] → BPGSP[t] + 2 Phos[t] },

(*Pentose phosphate pathway reactions.*)

{g6pdh,   Glc6P[t] + NADP[t] → P6GL[t] + NADPH[t] },

{pglhydrolysis, P6GL[t] → P6G[t] },

{p6gdh,

 P6G[t] + NADP[t] → CO2[t] + Ru5P[t] + NADPH[t]},

{gssgr,   GSSG[t] + NADPH[t] → 2 GSH[t] + NADP[t] },

{ru5pe,   Ru5P[t] → Xu5P[t]},

{r5pi,   Ru5P[t] → Rib5P[t]},

{tk1,   TK[t] + Xu5P[t] → TK$Xu5P[t]},

{tk2,   TK$Xu5P[t] → TKG[t] + GraP[t]},

{tk3,   TKG[t] + Rib5P[t] → TKG$Rib5P[t]},

{tk4,   TKG$Rib5P[t] → TK[t] + Sed7P[t]},

{tk5,   TKG[t] + Ery4P[t] → TKG$Ery4P[t] },

{tk6,   TKG$Ery4P[t] → TK[t] + Fru6P[t] },

{ta,   Sed7P[t] + GraP[t] → Ery4P[t] + Fru6P[t]},

(*Energy Consumption and oxidative reactions.*)

{ak,   MgADP[t] + ADP[t] → MgATP[t] + AMP[t]},

{atpase, MgATP[t] → MgADP[t] + Phos[t]},

{ox,   2 GSH[t] → GSSG[t] },

{oxnadh, NADH[t] → NAD[t]},

(*Membrane transport.*)

$\left\{ \text{lactransport}, \quad \dfrac{1}{\text{Vol}_i} \text{Lac}[t] \rightarrow \dfrac{1}{\text{Vol}_e} \text{Lace}[t] \right\}$,

$$\{\text{pyrtransport}, \quad \frac{1}{\text{Vol}_i} \text{ Pyr[t]} \rightarrow \frac{1}{\text{Vol}_e} \text{ Pyre[t]}\},$$

$$\{\text{phostransport}, \quad \frac{1}{\text{Vol}_i} \text{ Phos[t]} \rightarrow \frac{1}{\text{Vol}_e} \text{ Phose[t]}\},$$

```
(*Mg-metabolite binding.*)
{mgatp,    Mg[t] + ATP[t] → MgATP[t]},
{mgadp,    Mg[t] + ADP[t] → MgADP[t]},
{mgb23pg,  Mg[t] + B23PG[t] → Mg$B23PG[t]},
{mgb13pg,  Mg[t] + B13PG[t] → Mg$B13PG[t]},
{mgfru16p2, Mg[t] + Fru16P2[t] → Mg$Fru16P2[t]},
{mgglc16p2, Mg[t] + Glc16P2[t] → Mg$Glc16P2[t]},
{mgphos,   Mg[t] + Phos[t] → Mg$Phos[t]},

(*Hb-metabolite binding.*)
{hbmgatp,  Hb[t] + MgATP[t] → Hb$MgATP[t]},
{hbatp,    Hb[t] + ATP[t] → Hb$ATP[t]},
{hbadp,    Hb[t] + ADP[t] → Hb$ADP[t]},
{hbbpg,    Hb[t] + B23PG[t] → Hb$B23PG[t]},
{hbb13pg,  Hb[t] + B13PG[t] → Hb$B13PG[t]}
};
```

By using the previously mentioned trick, most equations in the list are written as per usual. It is only the hexokinase reaction and the three membrane transport reactions that include volume terms. The hexokinase reaction requires volume terms because it implicitly involves transport since glucose is assumed not to 'see' the cell membrane and thus has a compartment volume equal to $\text{Vol}_e + \text{Vol}_i$. For these four reactions, the reaction rates have units of mol s^{-1}, while all other reactions have units of mol L^{-1} s^{-1}. When we use **NMatrix** on this list we will generate $V^{-1} N$ in one step.

Another important point about this list of reactions is that for two of them, 2,3-BPG synthase-phosphatase (BPGSP) and transaldolase (TA), the elementary steps of the enzymic reaction are included, rather than the overall stoichiometry of the reactions at a steady state. Obviously, those steps in which only the overall reaction stoichiometry is given are modelled with steady-state rate equations, while the enzymic reactions in which the elementary steps are given are modelled with a set of elementary rate equations.

The method used for modelling each enzymic reaction varies with the task at hand. For most applications, using a steady-state rate equation is simplest because it greatly reduces the number of differential equations in the model and reduces the "stiffness" of the set as well (see Chapter 1). In the above model two enzymic reactions are expressed in terms of the elementary rate equations because the particular enzymes catalyze a number of different reactions at a single active site. In this situation the form of the steady-state equation is very complicated and is unwieldy to modify if this is required.

Having defined the reaction list, we are in a position to generate $(\mathbf{V}^{-1}\,\mathbf{N})$. Before we do this, however, we must define the compartment volumes. In Section 4.8 we noted that the hematocrit of a sample gives the proportion of the sample that is cells. For our model we will take it to be 0.5 that is near to the usual hematocrit of whole blood. Also, the volume fraction of an erythrocyte of normal shape that is occupied by hemoglobin and the cytoskeleton is ~0.3. Hence, with these values we define the intracellular $(\mathbf{Vol_i})$ and extracellular $(\mathbf{Vol_e})$ volumes per L of erythrocyte suspension.

```
     7
α = ── (*Cell water fraction of total cell volume.*);
    10
     1
Ht = ─ (*Hematocrit.*);
     2
Vol_e = 1 - Ht;
Vol_i = α Ht;
```

Having defined the compartment volumes, we can now generate $(\mathbf{V}^{-1}\,\mathbf{N})$. We will call this matrix $\overline{\mathbf{VN}}$.

```
VN =
  NMatrix[eqns, {CO2[t], Glc[t], Lace[t], Phose[t], Pyre[t]}];
Dimensions[VN]

  {56, 53}
```

This input generates a stoichiometry matrix that is based on the assumption that the concentrations of CO_2, glucose, extracellular lactate, pyruvate, and phosphate are all constant. The latter is a reasonable assumption for the *in situ* erythrocyte where the concentrations of these metabolites are buffered in the blood plasma. The stoichiometry matrix is too large to print here but from the **Dimensions** function it is seen that it describes the stoichiometric relationships between 56 internal metabolites in 53 different reactions.

Q: Generate the substrate and reaction velocity vectors for the erythrocyte model.

A: Recall that to generate these vectors we require the **SMatrix** and **VMatrix** from the **MetabolicControlAnalysis** add-on package. Note also that as in the previous example we load in the definitions for each rate equation that are specified in the list by using the command **<<RBCequations**.

```
v̄ = VMatrix[eqns] ;
S̄ =
    SMatrix[eqns, {CO2[t], Glc[t], Lace[t], Phose[t], Pyre[t]}];
S̄

{ADP[t], AMP[t], ATP[t], B13PG[t], B23PG[t], BPGSP[t],
  BPGSPP[t], BPGSPP$P2GA[t], BPGSPP$P3GA[t], BPGSPP$Phos[t],
  BPGSP$B13PG[t], BPGSP$B23PG[t], Ery4P[t], Fru16P2[t],
  Fru6P[t], Glc16P2[t], Glc6P[t], GraP[t], GrnP[t], GSH[t],
  GSSG[t], Hb[t], Hb$ADP[t], Hb$ATP[t], Hb$B13PG[t],
  Hb$B23PG[t], Hb$MgATP[t], Lac[t], Mg[t], MgADP[t],
  MgATP[t], Mg$B13PG[t], Mg$B23PG[t], Mg$Fru16P2[t],
  Mg$Glc16P2[t], Mg$Phos[t], NAD[t], NADH[t], NADP[t],
  NADPH[t], P2GA[t], P3GA[t], P6G[t], P6GL[t], PEP[t],
  Phos[t], Pyr[t], Rib5P[t], Ru5P[t], Sed7P[t], TK[t],
  TKG[t], TKG$Ery4P[t], TKG$Rib5P[t], TK$Xu5P[t], Xu5P[t]}
```

7.4 *In Vivo* Steady State of the Erythrocyte

By defining the matrices and vectors S, $(V^{-1}N)$, and v, we have defined the model of the erythrocyte. An important question that can now be asked is "Does the model exhibit a steady state?" In Chapter 4, the function **SteadyState** was introduced; it allows us to answer this question. Unfortunately the present metabolic system is too large for the **SteadyState** algorithm to locate one. But in such situations a good first approach is to perform a simulation to determine if at least a quasi-steady state is attained. Then we can use the results of the simulation to provide an initial estimate of the steady state for use in the **NSteadyState** function.

Q: Simulate a time course of the model of metabolism of the human erythrocyte.

A: Before being able to perform the simulation the initial conditions must be specified. The initial values of metabolite concentrations to be used here are those we have measured on freshly extracted human red cells.[3] Even if the metabolites are ultimately partitioned between free, Hb-bound, and Mg-bound forms, it is assumed as a starting point that all metabolites are initially free in solution.

The so-called *external parameters* and their values to be in the model are as follows:

```
r[t] = 0.69 ; (*Donnan ratio.*)
k[+1] = 0.150 (*Intracellular K+ concentration.*);
pH1[t] = 7.2 (*Intracellular pH.*);
CO2[t] = 1.2 × 10⁻³;
```

```
Glc[t] = 5 × 10⁻³ ;
Lace[t] = 1.82 × 10⁻³ ;
Phose[t] = 1.92 × 10⁻³ ;
Pyre[t] = 85 × 10⁻⁶ ;
```

Next, for the initial conditions (concentrations) of the metabolites we use the values in the list **ic1**.

```
ic1 = {
    ADP[0] == 0.31 × 10⁻³,
    AMP[0] == 30 × 10⁻⁶,
    ATP[0] == 2.1 × 10⁻³,
    B13PG[0] == 0.7 × 10⁻⁶,
    B23PG[0] == 6.70 × 10⁻³,
    BPGSP[0] == 3.8 × 10⁻⁶,
    BPGSPP[0] == 0,
    BPGSPP$P2GA[0] == 0,
    BPGSPP$P3GA[0] == 0,
    BPGSPP$Phos[0] == 0,
    BPGSP$B13PG[0] == 0,
    BPGSP$B23PG[0] == 0,
    Ery4P[0] == 10 × 10⁻⁶,
    Fru16P2[0] == 2.7 × 10⁻⁶,
    Fru6P[0] == 13 × 10⁻⁶,
    Glc16P2[0] == 122 × 10⁻⁶,
    Glc6P[0] == 40 × 10⁻⁶,
    GraP[0] == 5.7 × 10⁻⁶,
    GrnP[0] == 19.0 × 10⁻⁶,
    GSH[0] == 3.2 × 10⁻³,
    GSSG[0] == 0.09 × 10⁻⁶,
    Hb[0] == 7 × 10⁻³,
    Hb$ADP[0] == 0,
    Hb$ATP[0] == 0,
    Hb$B13PG[0] == 0,
    Hb$B23PG[0] == 0,
    Hb$MgATP[0] == 0,
    Lac[0] == 1.4 * 10^-3,
    Mg[0] == 3.0 * 10^-3,
    MgADP[0] == 0,
    MgATP[0] == 0,
    Mg$B13PG[0] == 0,
    Mg$B23PG[0] == 0,
    Mg$Fru16P2[0] == 0,
    Mg$Glc16P2[0] == 0,
```

```
Mg$Phos[0] == 0,
NAD[0] == 60 × 10^-6,
NADH[0] == 0.14 × 10^-6,
NADP[0] == 0.125 × 10^-6,
NADPH[0] == 64 × 10^-6,
P2GA[0] == 10 × 10^-6,
P3GA[0] == 64 × 10^-6,
P6G[0] == 1.4 × 10^-7,
P6GL[0] == 1.4 × 10^-10,
PEP[0] == 23 × 10^-6,
Phos[0] == 1.0 × 10^-3,
Pyr[0] == 60 × 10^-6,
Rib5P[0] == 10 × 10^-6,
Ru5P[0] == 10 × 10^-6,
Sed7P[0] == 10 × 10^-6,
TK[0] == 3.3 × 10^-7,
TKG[0] == 0,
TKG$Ery4P[0] == 0,
TKG$Rib5P[0] == 0,
TK$Xu5P[0] == 0,
Xu5P[0] == 1 × 10^-6};
```

With the initial conditions specified a simulation can be performed. The system is allowed to evolve for 1×10^6 s to determine whether it converges on at least a quasi-steady state.

```
sol =
NDSolveMatrix[S̄, V̄N, v̄, ic1, {t, 0, 1×10^6}, AccuracyGoal → 10,
  PrecisionGoal → 10, WorkingPrecision → 15, MaxSteps → 2000];
```

Plots of total 2,3-BPG and Glc6P concentrations over this time course are generated as follows:

```
Plot[Evaluate[{Glc6P[t]} /. sol], {t, 0, 1*^6},
  PlotRange -> {{0, 1×10^6}, {0., 50.×10^-6}},
  AxesLabel -> {"Time", "Concentration (M)"}];
```

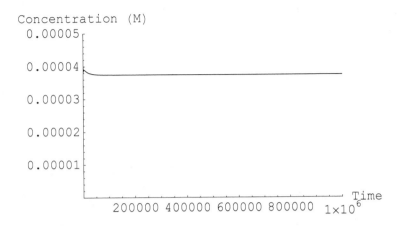

Figure 7.2. Simulated time course of glucose 6-phosphate concentration in a suspension of human erythrocytes.

```
Plot[Evaluate[{B23PG[t] + Hb$B23PG[t] + Mg$B23PG[t]} /. sol],
  {t, 0, 1×10⁶}, PlotRange -> {0., 8×10⁻³},
  AxesLabel -> {"Time", "Concentration (M)"}];
```

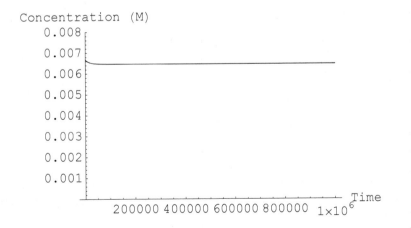

Figure 7.3. Simulated time course of 2,3-bisphosphoglycerate concentration in human erythrocytes.

From these representative time courses it appears that the system evolves to a steady state. Values of all metabolites at $t = 1 \times 10^6$ s (278 h) are

```
sValues = S̄ /. t → 1×10⁶ /. sol;
sTable = Transpose[{S̄, Flatten[sValues]}];
sTable
```

{{ADP[t], 0.00010835}, {AMP[t], 0.0000293475},
  {ATP[t], 0.000153333}, {B13PG[t], 3.57355×10⁻⁷},
  {B23PG[t], 0.00295352}, {BPGSP[t], 3.10247×10⁻⁷},
  {BPGSPP[t], 6.91289×10⁻⁷}, {BPGSPP$P2GA[t], 8.28742×10⁻¹²},
  {BPGSPP$P3GA[t], 9.07249×10⁻¹¹},
  {BPGSPP$Phos[t], 1.12746×10⁻⁶},
  {BPGSP$B13PG[t], 2.16381×10⁻⁸},
  {BPGSP$B23PG[t], 1.64938×10⁻⁶},
  {Ery4P[t], 7.09617×10⁻⁷}, {Fru16P2[t], 2.2274×10⁻⁶},
  {Fru6P[t], 0.0000121442}, {Glc16P2[t], 0.000121845},
  {Glc6P[t], 0.00003746}, {GraP[t], 5.14369×10⁻⁶},
  {GrnP[t], 0.000021374}, {GSH[t], 0.00320001},
  {GSSG[t], 8.6459×10⁻⁸}, {Hb[t], 0.00372286},
  {Hb$ADP[t], 0.000100843}, {Hb$ATP[t], 0.000205501},
  {Hb$B13PG[t], 4.21287×10⁻⁷}, {Hb$B23PG[t], 0.00274888},
  {Hb$MgATP[t], 0.000221496}, {Lac[t], 0.00140482},
  {Mg[t], 0.000383448}, {MgADP[t], 0.0000955809},
  {MgATP[t], 0.00152555}, {Mg$B13PG[t], 2.61032×10⁻⁸},
  {Mg$B23PG[t], 0.000760772}, {Mg$Fru16P2[t], 2.83559×10⁻⁹},
  {Mg$Glc16P2[t], 1.55114×10⁻⁷}, {Mg$Phos[t], 0.0000129705},
  {NAD[t], 0.0000598949}, {NADH[t], 2.45135×10⁻⁷},
  {NADP[t], 1.12787×10⁻⁷}, {NADPH[t], 0.0000640122},
  {P2GA[t], 0.0000118539}, {P3GA[t], 0.0000709407},
  {P6G[t], 0.0000322494}, {P6GL[t], 1.12795×10⁻⁸},
  {PEP[t], 0.0000198158}, {Phos[t], 0.000994883},
  {Pyr[t], 0.0000585718}, {Rib5P[t], 4.70849×10⁻⁶},
  {Ru5P[t], 3.99151×10⁻⁶}, {Sed7P[t], 5.53726×10⁻⁶},
  {TK[t], 1.49963×10⁻⁷}, {TKG[t], 1.7198×10⁻⁷},
  {TKG$Ery4P[t], 1.45191×10⁻⁹}, {TKG$Rib5P[t], 1.41216×10⁻⁹},
  {TK$Xu5P[t], 5.19321×10⁻⁹}, {Xu5P[t], 7.28302×10⁻⁶}}

These metabolite concentrations, along with the associated metabolic fluxes, are in good agreement with some experimental findings.[2] Note that a more visually friendly version of this table can be produced by using the **//MatrixForm** command.

Having obtained an initial estimate for the *in vivo* steady state, we now apply the **NSteadyState** function to refine this estimate. But before we do this we must determine the conservation of mass relationships that exist in the model of erythrocyte metabolism.

7.5 Conservation of Mass Relationships

From a knowledge of the stoichiometry matrix alone it is possible to determine the conservation relationships of the metabolic reaction scheme. This is done in the following question/answer.

Q: How do we determine the existence and nature of any conservation of mass relationships that might exist in the model of erythrocyte metabolism?

A: To answer this question, as was done in Chapter 4, we apply the function **ConservationRelations** from the **MetabolicControlAnalysis** package to the model.

> **ConservationRelations[$\bar{S}$, $\overline{VN}$]**

```
{ADP[t] + AMP[t] + ATP[t] + Hb$ADP[t] +
   Hb$ATP[t] + Hb$MgATP[t] + MgADP[t] + MgATP[t],
 BPGSP[t] + BPGSPP[t] + BPGSPP$P2GA[t] + BPGSPP$P3GA[t] +
 BPGSPP$Phos[t] + BPGSP$B13PG[t] + BPGSP$B23PG[t],
 Glc16P2[t] + Mg$Glc16P2[t], GSH[t] + 2 GSSG[t],
 Hb[t] + Hb$ADP[t] + Hb$ATP[t] +
   Hb$B13PG[t] + Hb$B23PG[t] + Hb$MgATP[t],
 Hb$MgATP[t] + Mg[t] + MgADP[t] + MgATP[t] + Mg$B13PG[t] +
  Mg$B23PG[t] + Mg$Fru16P2[t] + Mg$Glc16P2[t] + Mg$Phos[t],
 NAD[t] + NADH[t], NADP[t] + NADPH[t],
 TK[t] + TKG[t] + TKG$Ery4P[t] + TKG$Rib5P[t] + TK$Xu5P[t]}
== {Const[1], Const[2], Const[3], Const[4],
Const[5], Const[6], Const[7], Const[8], Const[9]}
```

Hence in the above model there are conservations of mass of (1) the adenosine moiety; (2) BPGSP (3) Glc(1,6)P2; (4) glutathione; (5)   hemoglobin; (6) magnesium; (7) NAD(P)(H); and (8) transketolase.

When performing the numerical simulation in Section 7.4 we did not use the conservation relationships to reduce the dimensions of the system of differential equations. Although from a consideration of the stoichiometry the above conservation relationships must apply, they will not automatically hold for the simulation unless the numerical integration is very accurate. Thus, one test of the accuracy of a simulation is to test the continued validity of these relationships throughout the simulated time course. This is illustrated next.

```
Plot[Evaluate[ADP[t] + AMP[t] + ATP[t] + Hb$ADP[t] +
    Hb$ATP[t] + Hb$MgATP[t] + MgADP[t] + MgATP[t] /. sol],
  {t, 0, 1×10⁶}, PlotRange → {0.0023, 0.0025},
  AxesLabel -> {"Time", "Concentration (M)"}];
```

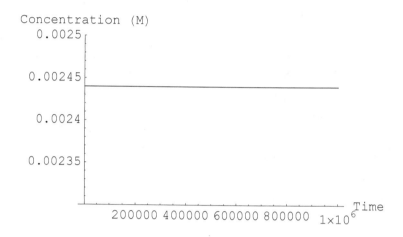

Figure 7.4. Checking that the conservation relationship between adenosine moieties holds for the simulated time course.

Figure 7.4 shows that the sum of the concentrations of adenosine moieties is constant during the ~280 h time course. Having determined that all eight conservation of mass relationships hold we can improve the estimate of the *in vivo* steady state of the metabolism by using the function **NSteadyState** from the **MetabolicControl-Analysis** package.

Q: Use the function **NSteadyState** to estimate the values of concentrations attained in the simulation presented in Section 7.4.

A: Before applying **NSteadyState** it is necessary to assign values to the constants of the conservation of mass relationships. These constants can be determined by evaluating them after using the initial conditions of the *in vivo* system. This can be done by cutting-and-pasting the output of the question/answer in Section 7.4, as follows:

```
{Const[1], Const[2], Const[3], Const[4],
  Const[5], Const[6], Const[7], Const[8], Const[9]} =
  {ADP[t] + AMP[t] + ATP[t] + Hb$ADP[t] + Hb$ATP[t] + Hb$MgATP[t] +
      MgADP[t] + MgATP[t], BPGSP[t] + BPGSPP[t] + BPGSPP$P2GA[t] +
      BPGSPP$P3GA[t] + BPGSPP$Phos[t] + BPGSP$B13PG[t] +
      BPGSP$B23PG[t], Glc16P2[t] + Mg$Glc16P2[t],
      GSH[t] + 2 GSSG[t], Hb[t] + Hb$ADP[t] + Hb$ATP[t] +
```

```
    Hb$B13PG[t] + Hb$B23PG[t] + Hb$MgATP[t],
  Hb$MgATP[t] + Mg[t] + MgADP[t] + MgATP[t] + Mg$B13PG[t] +
    Mg$B23PG[t] + Mg$Fru16P2[t] + Mg$Glc16P2[t] + Mg$Phos[t],
  NAD[t] + NADH[t], NADP[t] + NADPH[t], TK[t] + TKG[t] +
    TKG$Ery4P[t] + TKG$Rib5P[t] + TK$Xu5P[t]} /. t → 0 /. sol[[1]]
```

$\{0.00244, 3.8 \times 10^{-6}, 0.000122, 0.00320018,$
$0.007, 0.003, 0.00006014, 0.000064125, 3.3 \times 10^{-7}\}$

It is necessary to convert the estimate of the steady-state concentrations, that are obtained via the simulation, into the appropriate replacement rule. This is done as follows:

```
sValues = S̄ /. t → 1×10⁶ /. sol;
initialEst = Table[S̄[[i]] -> sValues[[1, i]], {i, Length[S̄]}]
```

$\{ADP[t] \rightarrow 0.00010835, AMP[t] \rightarrow 0.0000293475,$
$ATP[t] \rightarrow 0.000153333, B13PG[t] \rightarrow 3.57355 \times 10^{-7},$
$B23PG[t] \rightarrow 0.00295352, BPGSP[t] \rightarrow 3.10247 \times 10^{-7},$
$BPGSPP[t] \rightarrow 6.91289 \times 10^{-7}, BPGSPP\$P2GA[t] \rightarrow 8.28742 \times 10^{-12},$
$BPGSPP\$P3GA[t] \rightarrow 9.07249 \times 10^{-11}, BPGSPP\$Phos[t] \rightarrow 1.12746 \times 10^{-6},$
$BPGSP\$B13PG[t] \rightarrow 2.16381 \times 10^{-8}, BPGSP\$B23PG[t] \rightarrow 1.64938 \times 10^{-6},$
$Ery4P[t] \rightarrow 7.09617 \times 10^{-7}, Fru16P2[t] \rightarrow 2.2274 \times 10^{-6},$
$Fru6P[t] \rightarrow 0.0000121442, Glc16P2[t] \rightarrow 0.000121845,$
$Glc6P[t] \rightarrow 0.00003746, GraP[t] \rightarrow 5.14369 \times 10^{-6},$
$GrnP[t] \rightarrow 0.000021374, GSH[t] \rightarrow 0.00320001,$
$GSSG[t] \rightarrow 8.6459 \times 10^{-8}, Hb[t] \rightarrow 0.00372286,$
$Hb\$ADP[t] \rightarrow 0.000100843, Hb\$ATP[t] \rightarrow 0.000205501,$
$Hb\$B13PG[t] \rightarrow 4.21287 \times 10^{-7}, Hb\$B23PG[t] \rightarrow 0.00274888,$
$Hb\$MgATP[t] \rightarrow 0.000221496, Lac[t] \rightarrow 0.00140482,$
$Mg[t] \rightarrow 0.000383448, MgADP[t] \rightarrow 0.0000955809,$
$MgATP[t] \rightarrow 0.00152555, Mg\$B13PG[t] \rightarrow 2.61032 \times 10^{-8},$
$Mg\$B23PG[t] \rightarrow 0.000760772, Mg\$Fru16P2[t] \rightarrow 2.83559 \times 10^{-9},$
$Mg\$Glc16P2[t] \rightarrow 1.55114 \times 10^{-7}, Mg\$Phos[t] \rightarrow 0.0000129705,$
$NAD[t] \rightarrow 0.0000598949, NADH[t] \rightarrow 2.45135 \times 10^{-7},$
$NADP[t] \rightarrow 1.12787 \times 10^{-7}, NADPH[t] \rightarrow 0.0000640122,$
$P2GA[t] \rightarrow 0.0000118539, P3GA[t] \rightarrow 0.0000709407,$
$P6G[t] \rightarrow 0.0000322494, P6GL[t] \rightarrow 1.12795 \times 10^{-8},$
$PEP[t] \rightarrow 0.0000198158, Phos[t] \rightarrow 0.000994883,$
$Pyr[t] \rightarrow 0.0000585718, Rib5P[t] \rightarrow 4.70849 \times 10^{-6},$
$Ru5P[t] \rightarrow 3.99151 \times 10^{-6}, Sed7P[t] \rightarrow 5.53726 \times 10^{-6},$
$TK[t] \rightarrow 1.49963 \times 10^{-7}, TKG[t] \rightarrow 1.7198 \times 10^{-7},$
$TKG\$Ery4P[t] \rightarrow 1.45191 \times 10^{-9}, TKG\$Rib5P[t] \rightarrow 1.41216 \times 10^{-9},$
$TK\$Xu5P[t] \rightarrow 5.19321 \times 10^{-9}, Xu5P[t] \rightarrow 7.28302 \times 10^{-6}\}$

The steady-state concentrations can then be estimated with the following function:

```
steadyState = NSteadyState[S̄, V̄N̄, v̄, initialEst]
```

{ADP[t] → 0.00010835, AMP[t] → 0.0000293472,
 ATP[t] → 0.000153334, B13PG[t] → 3.57358×10⁻⁷,
 B23PG[t] → 0.00295354, BPGSP[t] → 3.10236×10⁻⁷,
 BPGSPP[t] → 6.91269×10⁻⁷, BPGSPP$P2GA[t] → 8.28719×10⁻¹²,
 BPGSPP$P3GA[t] → 9.07224×10⁻¹¹, BPGSPP$Phos[t] → 1.12743×10⁻⁶,
 BPGSP$B13PG[t] → 2.16375×10⁻⁸, BPGSP$B23PG[t] → 1.64933×10⁻⁶,
 Ery4P[t] → 7.09623×10⁻⁷, Fru16P2[t] → 2.22743×10⁻⁶,
 Fru6P[t] → 0.0000121444, Glc16P2[t] → 0.000121845,
 Glc6P[t] → 0.0000374605, GraP[t] → 5.14373×10⁻⁶,
 GrnP[t] → 0.0000213741, GSH[t] → 0.00320001,
 GSSG[t] → 8.6459×10⁻⁸, Hb[t] → 0.00372285,
 Hb$ADP[t] → 0.000100842, Hb$ATP[t] → 0.000205502,
 Hb$B13PG[t] → 4.2129×10⁻⁷, Hb$B23PG[t] → 0.00274889,
 Hb$MgATP[t] → 0.000221496, Lac[t] → 0.00140482,
 Mg[t] → 0.000383447, MgADP[t] → 0.0000955801,
 MgATP[t] → 0.00152555, Mg$B13PG[t] → 2.61033×10⁻⁸,
 Mg$B23PG[t] → 0.000760774, Mg$Fru16P2[t] → 2.83561×10⁻⁹,
 Mg$Glc16P2[t] → 1.55114×10⁻⁷, Mg$Phos[t] → 0.0000129705,
 NAD[t] → 0.0000598949, NADH[t] → 2.45135×10⁻⁷,
 NADP[t] → 1.12786×10⁻⁷, NADPH[t] → 0.0000640122,
 P2GA[t] → 0.0000118539, P3GA[t] → 0.0000709407,
 P6G[t] → 0.0000322497, P6GL[t] → 1.12795×10⁻⁸,
 PEP[t] → 0.0000198158, Phos[t] → 0.000994883,
 Pyr[t] → 0.0000585718, Rib5P[t] → 4.70853×10⁻⁶,
 Ru5P[t] → 3.99154×10⁻⁶, Sed7P[t] → 5.53733×10⁻⁶,
 TK[t] → 1.49963×10⁻⁷, TKG[t] → 1.7198×10⁻⁷,
 TKG$Ery4P[t] → 1.45192×10⁻⁹, TKG$Rib5P[t] → 1.41217×10⁻⁹,
 TK$Xu5P[t] → 5.19324×10⁻⁹, Xu5P[t] → 7.28307×10⁻⁶}

Inspection of the above steady-state concentrations indicates that they are essentially the same as those determined by simulation of the time course for ~280 h. This can also be confirmed by the following calculation:

```
sss = S̄ /. steadyState;
sie = S̄ /. initialEst;
sie - sie
─────────
   sss
```

{0., 0., 0., 0., 0., 0., 0., 0., 0., 0., 0., 0., 0., 0.,
 0., 0., 0., 0., 0., 0., 0., 0., 0., 0., 0., 0., 0., 0.,
 0., 0., 0., 0., 0., 0., 0., 0., 0., 0., 0., 0., 0., 0.,
 0., 0., 0., 0., 0., 0., 0., 0., 0., 0., 0., 0.}

The calculation shows that the relative difference between the simulated steady-state concentrations and those calculated with the function **NSteadyState** from the **MetabolicControlAnalysis** add-on package is insignificant!

Q: Is the steady state that is calculated above stable?

A: To answer this question we use the **Stability** function from the **MetabolicControlAnalysis** package. The following *Mathematica* input is used, with the consequent diagnostic output.

Stability[$\bar{S}$, $\overline{VN}$, $\bar{v}$, SteadyStateConc → steadyState]

```
Asymptotically Unstable
```

```
{-1.00023×10⁸, -1.60544×10⁶, -19888.8, -6343.18, -4238.33,
 -3300.83, -3056.35, -2979.01, -2945.63, -1462.18,
 -1217.33, -1201.55, -1201.53, -1201.03, -1200., -1126.86,
 -424.786, -217.858, -215.172, -191.47, -103.524,
 -73.2418, -40.9648, -10.2955, -7.29762, -5.30389,
 -5.26565, -4.38785, -2.18217, -1.28346, -1.19732,
 -1.021, -0.821454, -0.413529, -0.30293, -0.207559,
 -0.148637, -0.0253603, -0.0159573 + 0.00485246 i,
 -0.0159573 - 0.00485246 i, -0.0146684, -0.00730836,
 -0.00601884, -0.00181841, -0.00130747, -0.00105352,
 -0.00005637, 8.8251×10⁻⁸, -1.20209×10⁻⁹, -1.12108×10⁻¹⁰,
 -1.19318×10⁻¹¹, 8.65173×10⁻¹⁴, 2.13345×10⁻¹⁴,
 1.80779×10⁻¹⁵, 2.05997×10⁻¹⁷, 6.26896×10⁻¹⁸}
```

The list of eigenvalues has three non-negative real numbers so the **Stability** function returned an "unstable" verdict. On the other hand, these values are very small and close to the precision limit of the calculations. Since all the negative real values are also effectively zero, establishing the existence of stability of the steady state is not possible with the present type of analysis (see Section 4.7).

However, it can be shown on repeated simulation that the same (quasi-)steady state is arrived at after long times from a wide range of initial conditions. This suggests that at least operationally the steady state is stable.

7.6 Simulating a Time Course

So far, we have examined the behavior of the erythrocyte model with the assumption that metabolites such as CO_2, glucose, extracellular lactate, pyruvate, and phosphate remain at constant fixed concentrations; in other words, these metabolites are treated as so-called *external* metabolites. The assumption of this situation is reasonable because *in situ* the erythrocyte is exposed to approximately constant concentrations in the blood plasma. This enables the erythrocyte to establish a steady state of metabolite concentrations in this thermodynamically open system. However, in many important experimental conditions and many disease states, erythrocyte metabolism does not attain a steady state and the concentrations of these metabolites can no longer be treated as fixed in a simulation.

We finish this chapter with an example that reveals how the present model of erythrocyte metabolism[2–4] simulates time-dependent changes that occur over short times before the attainment of a global steady state of metabolite concentrations. This type of analysis is very useful for the interpretation of experimental data as well as for testing the behavior of the model under conditions that simulate inborn errors of enzyme function.

Q: Simulate the time course of glucose, total 2,3-BPG, and lactate concentrations that might occur in an experiment in which erythrocytes (hematocrit = 0.5) are incubated with an initial concentration of 10 mmol L^{-1} glucose and assuming constant CO_2. Assume that during the time course the intracellular pH decreased linearly from 7.2 to 6.8 in 10 h.

A: The starting point for this simulation requires external parameters, initial conditions, and equations from the previous simulation. However, these are required to be modified to simulate the new time course.

First, the external parameters are identified and their concentrations specified. Four of the parameters that were defined as external ones in the previous model are now internal parameters. Hence we must **Clear** these values.

```
Glc[t] =.; Lace[t] =.; Phose[t] =.; Pyre[t] =.;
```

It is assumed that CO_2 remains as an external parameter (which has already been assigned a value of 1.2 mmol L^{-1}). In the model pH also remains an external parameter but its value will no longer be constant. Since pH is assumed to change linearly from 7.2 to 6.8 in 10 h, the following equation describes this change.

```
pH1[t] =.;
pH1[t_] := 7.2 - 0.4 * t / (36000);
```

Glucose, external lactate, pyruvate, and phosphate are no longer deemed to be external parameters, so initial conditions are required for them; these are appended to the initial conditions list, `ic1`, which was defined previously. This is achieved as follows:

```
ic2 = Union[{Glc[0] == 10 × 10⁻³,
             Lace[0] == 1.82 × 10⁻³,
             Phose[0] == 1.92 × 10⁻³,
             Pyre[0] == 85 × 10⁻⁶}, ic1];
```

All that remains to be done is to define the matrices and vectors of the new model. To do this, use is made of the original equation list, **eqns**, but CO_2 only is specified as an external parameter. Hence the new matrices and vectors are

```
n̄V̄N̄ = NMatrix[eqns, {CO2[t]}] ;
```

```
n̄S̄ = SMatrix[eqns, {CO2[t]}] ;
```

The reaction velocity vector remains unchanged from previously. Now simulate some selected time courses of concentrations.

```
sol2 = NDSolveMatrix[n̄S̄, n̄V̄N̄, v̄, ic2,
    {t, 0, 36000}, AccuracyGoal → 10, PrecisionGoal → 10,
    WorkingPrecision → 15, MaxSteps → 2000];
```

```
Plot[Evaluate[{Glc[t], (Lac[t] Volᵢ + Lac[t] Volₑ) / (Volₑ + Volᵢ)}
    /. sol2], {t, 0, 30000},
  AxesLabel -> {"Time", "Concentration (M)"}];
```

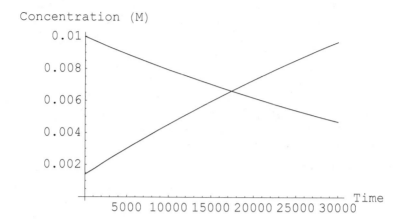

Figure 7.5. Simulated time course of glucose consumption (declining curve) and lactate production (increasing curve) for the model of human erythrocyte metabolism.

```
Plot[Evaluate[{B23PG[t] + Hb$B23PG[t] + Mg$B23PG[t]} /. sol2],
   {t, 0, 30000}, PlotRange -> {0., 8×10⁻³},
   AxesLabel -> {"Time", "Concentration (M)"}];
```

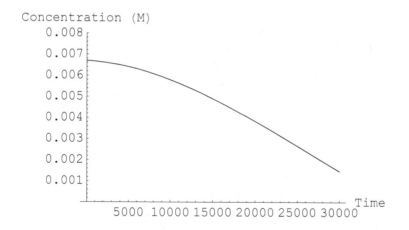

Figure 7.6. Simulated time course of 2,3-BPG depletion in human erythrocytes.

These simulations demonstrate that in such an experiment the concentration of total 2,3-BPG would be expected to decline from an initial concentration of ~7 mmol L^{-1} to a concentration of ~0 mmol L^{-1} after 10 h. In fact, when NMR spectroscopy was used to follow the metabolite changes in such an experiment this is precisely what was observed.[2] The decline in 2,3-BPG was shown to be mainly due to the decrease in pH that occurs during the experiment and the fact that a low pH inhibits many enzymes of glycolysis and the 2,3-BPG shunt. The inhibition of glycolytic enzymes is reflected in the plot of glucose and lactate concentrations during the experiment; as the pH drops, the rate of glucose utilization (and lactate production) falls.

The role of pH in regulating erythrocyte metabolism is addressed in greater detail in the next chapter, but for now an easy way to demonstrate its importance is to perform Exercise 7.6.

7.7 Exercises

7.7.1

Print out the stoichiometry matrix generated in Section 7.3 and check its validity for several of the more obvious reactants, such as Fru6P and 2,3-BPG.

7.7.2

Determine the flux through hexokinase at $t = 1,000,000$ s in the simulation performed in Section 7.4. Define the units of the flux.

7.7.3

Test that *all* the conservation of mass relationships are adhered to during the simulation performed in Section 7.4.

7.7.4

Repeat the question/answer in Section 7.8; however, on this occasion assume that intracellular pH is constant throughout the experiment.

7.7.5

In Sections 7.4 and 7.5 the normal *in vivo* steady states of metabolite concentrations in the erythrocyte were determined. What is the rate of glycolysis under the steady-state condition? What happens to this rate of glycolysis if the total magnesium concentration in the erythrocyte is reduced by 50%?

7.7.6

What happens to the steady-state concentration of 2,3-BPG when intracellular pH is increased to 7.3 from 7.2? How long does it take for [2,3-BPG] to be halfway between the normal and the new steady-state value?

7.8 References

1. Mulquiney, P.J. and Kuchel, P.W. (1997) Model of the pH-dependence of the concentrations of complexes involving metabolites, haemoglobin and magnesium ions in the human erythrocyte. *Eur. J. Biochem.* **245**, 71-83.
2. Mulquiney, P.J., Bubb, W.A., and Kuchel, P.W. (1999a) Model of 2,3-bisphosphoglycerate metabolism in the human erythrocyte based on detailed enzyme kinetic equations: *in vivo* kinetic characterisation of 2,3-bisphosphoglycerate synthase/phosphatase using ^{13} C and ^{31} P NMR. *Biochem. J.* **342**, 567-580.
3. Mulquiney, P.J. and Kuchel, P.W. (1999b) Model of 2,3-bisphosphoglycerate metabolism in the human erythrocyte based on detailed enzyme kinetic equations: computer simulation and metabolic control analysis. *Biochem. J.* **342**, 597-604.
4. Mulquiney, P.J. and Kuchel, P.W. (1999c) Model of 2,3-bisphosphoglycerate metabolism in the human erythrocyte based on detailed enzyme kinetic equations: equations and parameter refinement. *Biochem. J.* **342**, 581-596.
5. Heinrich, R. and Rapoport, T.A. (1973) A linear steady-state treatment of enzymatic chains. General properties, control and effector strength. *Eur. Biochem. J.* **42**, 89-95.
6. Heinrich, R. and Rapoport, T.A. (1973) A linear steady-state treatment of enzymatic chains. Critique of the cross-over theorem and a general procedure to identify interaction sites with an effector. *Eur. Biochem. J.* **42**, 97-105.
7. Heinrich, R., Rapoport, S.M., and Rapoport, T.A. (1977) Metabolic regulation and mathematical models. *Prog. Biophys. Mol. Biol.* **32**, 1-82.

8 Metabolic Control Analysis of Human Erythrocyte Metabolism

8.1 Introduction

The simulations that were presented in Chapter 7 allowed a comparison of the behavior of the model of erythrocyte metabolism with real experimental results (see the original papers[1-4] for the data). While this was an extremely important step in model verification, it left many unanswered questions about the underlying regulatory and control features. In Chapter 5 we introduced the concepts of MCA and described how they provide greater insights into a model's behavior than would be unattainable with only descriptive language. In this chapter we apply the procedures of MCA to our model.

The main questions that are addressed are: Which reactions are important for controlling pathway fluxes? Which reactions are important for controlling metabolite concentrations? What is the mechanism for the control of 2,3-BPG concentration?

8.2 Normal *In Vivo* Steady State

In the following sections the various coefficients of MCA are calculated. These coefficients are defined with respect to a particular steady state of the system. We also ask questions about the control and regulation of erythrocyte metabolism as it pertains in the neighborhood of the normal *in vivo* steady state of metabolite concentrations. The steady-state values used in the following discussions are those that were simulated in Chapter 7.

Q: Calculate the normal *in vivo* steady state of the erythrocyte for the model described in Chapter 7.

A: The simulation method of Chapter 7 is used to calculate the concentration of metabolites in the normal *in vivo* steady state. To use this method we need to load in the red blood cell equations and the metabolic control analysis package. We also must re-enter the equation list for the model and the values of any external parameters and initial conditions. The equation list and the external parameters and initial conditions of the erythrocyte model are found in Appendices 4 and 5. By evaluating the cells in these appendices, we will then be able to load the equation list and the values with the **<<** command.

All these tasks are performed with the following commands:

```
<< RBCequations;
<< eqns;
<< initialconditions;
<< MetabolicControlAnalysis`
```

We are now in a position to define the matrices and vectors for the model. This is done as follows:

$\overline{VN}$ =
```
NMatrix[eqns, {CO2[t], Glc[t], Lace[t], Phose[t], Pyre[t]}] ;
```

$\bar{S}$ =
```
SMatrix[eqns, {CO2[t], Glc[t], Lace[t], Phose[t], Pyre[t]}];
```

$\bar{v}$:= VMatrix[eqns]

Next, use **NDSolveMatrix** as was done in Chapter 7.

```
sol =
    NDSolveMatrix[S̄, VN̄, v̄, icl, {t, 0, 1×10⁶}, AccuracyGoal → 10,
        PrecisionGoal → 10, WorkingPrecision → 15, MaxSteps → 2000] ;
```

A replacement rule to enable the insertion of the steady-state concentrations is defined by using the substrate values after 1×10^6 s of simulation.

```
sValues = S̄ /. t -> 1×10⁶ /. sol // Flatten;

steadyState = Table[S̄[[i]] → sValues[[i]], {i, Length[S̄]}];
```

The Table **steadyState** contains the concentrations of all the reactants. It can be inspected by re-running the Cell after deleting the semicolon at the end of the line.

8.3 Identifying Zero Fluxes

In the following sections normalized metabolic control coefficients are used in preference to non-normalized ones because they convey a better biochemical understanding as the result of having a (usually) well-defined range of values, $0 - 1$.

However, the normalization of coefficients involves division by the normalizing value which may be zero, thus rendering the coefficient undefinable. In other words, the normalized flux coefficients become undefined in situations when the flux is zero in a given steady state, and a concentration coefficient does likewise when the steady-state concentration is zero.

In the analysis above the pathway fluxes are calculated by using numerical methods, so there is no certainty whether a particular pathway flux is actually zero or simply close to it. Because there is a limit to the precision of any numerically calculated number, we can never be sure whether some numbers are simply very small or whether they are actually zero. Clearly, this has implications for calculating control coefficients.

Hence, an important question is: Are any of the fluxes in the erythrocyte model necessarily zero in the steady state? An efficient analytical procedure for answering this question is described in the following question/answer.

Q: Identify all zero fluxes in the mathematical model of erythrocyte metabolism at its steady state.

A: First, determine by simple inspection which reaction velocities are very small after 1×10^6 s of simulation. This is done as follows:

```
velocityTable = v̄ /. steadyState;

zeroFluxPosition =
  Position[velocityTable, x_ ? (Abs[#1] < 1×10⁻¹⁰ &)]

{{15}, {16}, {17}, {18}, {19}, {35}, {41}, {42}, {43}, {44},
  {45}, {46}, {47}, {48}, {49}, {50}, {51}, {52}, {53}}
```

The second line of the input returns the position in the list of those reaction velocities that are less than 1×10^{-10} mol L^{-1} s^{-1}.

Thus it appears that there are many reactions that have rates around 10^{-10} mol L^{-1} s^{-1}. The important question, from the point of view of calculating normalized control coefficients, is whether they really are zero or simply close to zero. This question is considered in the next question/answer.

Q: Is it possible to decide whether any of the reaction rates identified in the previous question/answer would actually be zero if the steady-state concentrations were known exactly?

A: It turns out that by analyzing the stoichiometry of the system, for some of these reactions, we can obtain the relevant information. The following functions pick out those fluxes which are necessarily zero in the steady state.

First, we calculate the so-called null space of the stoichiometry matrix[5] as follows:

```
null = Transpose[NullSpace[VN]] ;
```

Then by identifying those columns of the null space matrix which contain all zero elements, we can identify the reaction rates that are *necessarily* zero in the steady state.

```
zeroFluxPosition =
Position[null, Table[0, {i, Length[Transpose[null]]}]]
```

```
{{19}, {35}, {41}, {42}, {43}, {44}, {45},
 {46}, {47}, {48}, {49}, {50}, {51}, {52}, {53}}
```

An explanation of this method is given below.

In the method used above, the columns of the **null** matrix represent a basis set of vectors for the matrix equation[5]

$$\mathbf{N.null} = 0 \ . \tag{8.1}$$

This means that any solution of the equation

$$\mathbf{N \ v} = 0 \tag{8.2}$$

must be a linear combination of these basis vectors. The steady-state fluxes, **J**, are determined by solving Eqn [8.1]. Hence if any row of **null** consists entirely of zeros, then the flux in the same row in **J** will necessarily be 0. These conclusions are reached entirely by considering reaction stoichiometries; in other words, they are not affected by the values of any rate parameters in the system.

The reactions identified in the above example are in strict 'detailed balance' in every steady state of the system. Thus, the flux through each of these reactions cannot be influenced by any other reaction of the system (since their net flux is always zero). Therefore the non-normalized control coefficients expressing the control of other reactions on the flux through each strictly balanced reaction is zero; and the corresponding normalized MCA coefficient will be undefined. Therefore, to avoid problems when calculating control coefficients it is prudent to remove the zero-flux reactions from the matrices; furthermore, no valuable information is lost. This removal is done by the function **FluxControlMatrix** (see below) by using the option **Normalized** → **zerofluxposition**, where **zerofluxposition** is a list $\{\{i\},\{j\}, ...\}$ specifying the fluxes/reaction rates in rows/columns $i, j, ...$ which are necessarily zero.

8.4 Flux Control Coefficients

As is noted in Chapter 5, flux control coefficients give information on the relative importance of a particular reaction in controlling a particular flux in a metabolic pathway. Often, the easiest way to calculate the flux control coefficient for a model, that is formulated using matrix notation, is with the function **FluxControlMatrix** from the **MetabolicControlAnalysis** add-on package.

Q: Calculate the non-normalized flux control coefficient values for the erythrocyte model.

A: The matrix is calculated as follows:

```
fcm = FluxControlMatrix[S̄, V̄N̄, v̄,
    SteadyStateConc → steadyState, Normalized → False];
Dimensions[
  fcm]
```

```
Inverse::luc : Result for Inverse of badly conditioned
    matrix ≪1≫ may contain significant numerical errors.
```

{53, 53}

The resultant flux control matrix is a 53 × 53 one, since there are 53 reactions in the system and there is one possible flux associated with each reaction.

The error message that is returned when the previous function is evaluated needs some explanation. Inverse::luc sometimes appears when the **FluxControlMatrix** and **ConcControlMatrix** functions are used. These commands require the calculation of the inverse of the Jacobian (**MMatrix**) of the system. In some cases the Jacobian matrix is ill-conditioned, meaning that small perturbations in its elements cause large changes in the calculated inverse. In these cases, because the Jacobian is a numerical one, there may be significant errors in the calculated results. In practice, although this problem is something to be aware of, in most cases the results of the ControlMatrix algorithms closely match control coefficients calculated by numerical modulation (Section 5.3). This point is demonstrated in the following example, but we first switch off the error messages facility, as follows:

```
Off[Inverse::"luc"];
```

Q: Calculate the normalized flux control coefficients for the erythrocyte model.

A: From the discussion in Section 8.3 it is apparent that we must specify those reactions and fluxes that obey strict detailed-balance. As is noted above this is done using the Option **Normalized → zeroFluxPosition**.

```
fcm = FluxControlMatrix[S̄, V̄N̄, v̄, SteadyStateConc → steadyState,
    Normalized → zeroFluxPosition];

Dimensions[fcm]
```

{38, 53}

The use and interpretation of the flux control matrix is demonstrated by addressing the questions in the remainder of this section.

Q: How much control does hexokinase exert over the flux through glycolysis?

A: Before answering this question, recall that the flux control matrix has the following meaning: the entry in row i and column j gives the flux control coefficient of reaction j with respect to pathway flux i. In other words, this entry gives a measure of how reaction j controls the pathway flux through reaction i. Thus the flux control matrix will have dimensions of 38×53 because 53 reactions make up the model, but from Section 8.4 it is clear there are 15 equations that obey strict detailed-balance.

If the flux control coefficient of hexokinase, with respect to the flux through glycolysis (which we will also take as the pathway flux through hexokinase), is sought, then the matrix entry in row 1, column 1 of the flux control matrix must be inspected. The relevant flux control coefficient is given by the following evaluation. (The other flux control coefficients are treated in more examples below.)

```
cc = fcm[[1,1]]
```

```
0.00416549
```

From the flux-summation theorem of MCA (Section 5.2) we know that all the control coefficients for flux through glycolysis should sum to 1. Therefore, we can infer that hexokinase does not exhibit very much control over glycolytic flux in the normal steady state.

Q: Can the conclusion reached in the previous example be right? This is a surprising result in view of the claims in the earlier literature on the control of human erythrocyte metabolism.

A: In answering this question, we first determine if the flux-summation theorem holds for all the flux control coefficients. We do this by summing the entries for all the rows of the flux control matrix. This is done with the following input:

```
Table[Apply[Plus, fcm[[i]] ], {i, Length[fcm]}]
```

```
{1., 1., 1., 1., 1., 1., 1., 1., 1., 1., 1., 1., 1., 1., 1.00054,
 0.999997, 1.00054, 0.999997, 1., 1., 1., 1., 1., 1., 1.,
 1., 1., 1., 1., 1., 1., 1., 1., 1., 1., 1., 1.00002}
```

It is clear that each set of flux control coefficients sum to one, or very close to one. This provides consistency with the summation theorem for flux control coefficients (Section 5.2).

Q: In view of the results from the previous two question/answers, in which reactions is most of the control of glycolytic flux vested?

A: One way of answering this question is to determine which reactions have a flux control coefficient whose absolute value is greater than, say, 0.1. The following program implements this idea. It defines a function which returns the reaction name of any flux control coefficients which is greater than a specified value. The first step of the program is to define a list of reactions and a list of fluxes. The list of fluxes is simply the list of reactions (**reactionList**) with those reactions which have zero flux removed.

```
reactionList = Drop[StoichiometryMatrix[eqns][[1]], 1];
fluxList = Delete[reactionList, zeroFluxPosition];
```

We then generate replacement rules of the form column/row number → reaction/flux name and vice versa. These allow us to identify various rows and columns in the flux control matrix with particular fluxes and reactions.

```
numToRxn =
   Table[i → reactionList[[i]], {i, Length[reactionList]}];
rxnToNum = Table[reactionList[[i]] → i,
   {i, Length[reactionList]}];
numToFlux = Table[i → reactionList[[i]],
   {i, Length[reactionList]}];
fluxToNum = Table[reactionList[[i]] → i,
   {i, Length[reactionList]}];
```

This allows us to define a function, **CJ[x,y]**, which gives the control coefficients for flux x which are ≥ y.

```
CJ[x_, y_] := Module[
   {bigFccPosition, names, values},
   bigFccPosition =
     Position[fcm[[x/.fluxToNum]], z_ ? (Abs[#1] >= y &)] // Flatten;
   names = bigFccPosition /. numToRxn ;
   values = Part[fcm[[x/.fluxToNum]], bigFccPosition];
   Transpose[{names, values}]]
```

Thus, using the above definitions, the reactions that have an absolute value for their glycolytic flux control coefficient greater than 0.1 are given by

```
CJ[hk, 0.1]
```

```
{{bpgsp9, 0.181721}, {atpase, 0.679922}}
```

From the list it is evident that the reactions with the two largest glycolytic flux control coefficients are ATPase and the ninth reaction step of the 2,3-BPG synthase-phosphatase reaction scheme. This is a very interesting result!

Q: Which reactions are primarily responsible for controlling the flux through the pentose phosphate pathway (PPP)?

A: In answering this question we take the flux through the PPP to be flux via G6PDH. Hence, the reactions with the largest flux control coefficients can be determined as follows using the function **CJ** defined above:

```
CJ[g6pdh, 0.1]
```

```
{{ox, 0.940131}}
```

From this output we see that the reaction (ox) has the highest flux control coefficient with a value of 0.940131. Therefore, the answer is that the rate of oxidation of glutathione is primarily what controls PPP flux under the normal *in vivo* steady-state condition.

Q: Which reactions are primarily responsible for controlling the flux through the 2,3-BPG shunt?

A: In answering this question we take the flux through the second step of the 2,3-BPGsynthase-phophatase reaction scheme as the pathway flux of the 2,3-BPG shunt. Hence, the largest flux control coefficients can be determined as follows, using the function **CJ** defined above:

```
CJ[bpgsp2, 0.1]
```

```
{{eno, 0.108495}, {pk, 0.158887}, {bpgsp2, 0.1307},
 {bpgsp9, 0.836173}, {atpase, -0.124067}}
```

Therefore, the answer to the question is that the energetic and oxidative loads, almost independently, control the fluxes through glycolysis and the PPP, respectively, while the irreversible phosphatase reaction of the 2,3-BPG synthase-phosphatase enzyme has the greatest control over 2,3-BPG shunt activity.

Q: Calculate the flux control coefficient of hexokinase, with respect to the flux through glycolysis by numerical modulation. This analysis will serve as a check on the matrix calculations performed in the previous question/answers.

A: Recall from Chapter 4 that the numerical modulation method involves varying parameter values by a small, incremental amount as a numerical means of determining a partial derivative. For the hexokinase reaction this is done as follows.

Define **solution1** as the normal steady state of metabolite concentrations, as determined previously.

```
solution1 =
   NDSolveMatrix[S̄, V̄N̄, v̄, ic1, {t, 0, 1*^6}, AccuracyGoal → 10,
      PrecisionGoal → 10, WorkingPrecision → 15, MaxSteps → 2000];
```

The velocity of hexokinase at the steady state is then given by

```
v1 = v[hk] /. t → 1*^6 /. solution1;
```

Next we numerically modulate the hexokinase reaction. This is most easily achieved by increasing the concentration of hexokinase by, say, 1%.

```
HKold = HK;
HK = HK * 1.01;
```

Now determine the new steady-state velocity of hexokinase after the parameter modulation. But, before this is done a new reaction velocity matrix must be defined, as follows:

```
v̄2̄ := VMatrix[eqns] ;
```

```
solution2 =
   NDSolveMatrix[S̄, V̄N̄, v̄2̄, ic1, {t, 0, 1*^6}, AccuracyGoal → 10,
      PrecisionGoal → 10, WorkingPrecision → 15, MaxSteps → 2000];
```

```
v2 = v[hk] /. t → 1*^6 /. solution2;
```

Finally, the estimated flux control coefficient is determined as follows:

$$\text{fluxControlCoeff} = \frac{v2 - v1}{v1} \times \frac{1}{0.01}$$

```
HK = HKold;
```

```
{0.00306096}
```

In conclusion, this value agrees to closely with that calculated above using the matrix method!

8.5 Concentration Control Coefficients

The function `ConcentrationControlMatrix` in the `MetabolicControl-Analysis` add-on package can be used to calculate the concentration control coefficients for the erythrocyte model. These coefficient values express the importance of a particular reaction in controlling the concentration of a particular metabolite (Section 5.2).

Q: How is the concentration control coefficient matrix computed?

A: This is done by executing the following function with the three arguments that define the whole erythrocyte model, namely, $\bar{S}$, $\overline{VN}$, and $\bar{v}$:

```
ccm =
    ConcControlMatrix[S̄, V̄N̄, v̄, SteadyStateConc → steadyState];
Dimensions[ccm]

{56, 53}
```

The whole matrix is not printed here because of space restrictions, but it can be inspected by simply deleting the semicolon after the function and then re-evaluating the *Mathematica* Cell.

Also, note that each entry in row i and column j of the control matrix gives the concentration control coefficient of reaction j with respect to metabolite i. The concentration control coefficient matrix has dimensions of 56×53 because there are 56 internal substrate variables and 53 reactions in the erythrocyte model.

Q: Which reactions are important in controlling the concentration of 2,3-BPG?

A: Before answering this question directly, it turns out to be useful to define first some replacement rules and functions which make the overall analysis much simpler.

First define a replacement rule of the form substrate name → row, and one of the form row → substrate name.

```
substrateToNum = Table[S̄[[i]] → i, {i, Length[S̄]}];
numToSubstrate = Table[i → S̄[[i]], {i, Length[S̄]}];
```

Then, as was done for the flux control coefficients, we define a function, `CC[x,y]`, which gives the concentration control coefficients for concentration x which are $\geq$ y.

```
CC[x_ , y_] := Module[{bigCccPosition, names, values},
  bigCccPosition =
    Position[ccm⟦x/.substrateToNum⟧ , z_ ? (Abs[#1] >= y &)] // Flatten;
  names = bigCccPosition /. numToRxn ;
  values = Part[ccm⟦x/.substrateToNum⟧ , bigCccPosition];
  Transpose[{names, values}]]
```

Now check whether the summation theorem (Eqn [5.7]) holds for the control of 2,3-BPG concentration.

```
Apply[Plus, ccm[[B23PG[t] /. substrateToNum]]]
```

-1.90026×10^{-7}

Clearly, this output indicates a value that is very nearly 0, as expected, thus indicating a successful analysis. And finally, the important reactions for controlling free 2,3-BPG concentration are determined with the function **CC**, as follows:

```
CC[B23PG[t], 0.01]
```

```
{{hk, 1.57071}, {pfk, 0.345435}, {eno, -0.183782},
 {pk, -0.269142}, {bpgsp2, 0.127866}, {bpgsp9, -0.410397},
 {atpase, -1.16271}, {ox, -0.0190557}}
```

The list indicates that there are several enzymes that exert significant control over the concentration of 2,3-BPG. The most important of these are hexokinase, phosphofructokinase, enolase, pyruvate kinase, the two irreversible steps of the 2,3-BPG synthase-phosphatase, and the ATPase.

8.6 Response Coefficients and Partitioned Responses

Another class of question that is important to answer in relation to the behavior of a metabolic pathways is: How does the model respond to different external effectors? For example, experimental work on 2,3-BPG metabolism has shown that its concentration is very sensitive to changes in intracellular pH.[2] One way of characterizing this sensitivity is through the concentration response coefficient.

Q: What is the value of the response coefficient of 2,3-BPG concentration with respect to intracellular pH?

A: In order to calculate response and elasticity coefficients with respect to `pH1[t]` we must clear its set value and then reassign a value using a parameter matrix, as shown in Chapter 4. In addition, because the matrix of reaction velocities was determined with `pH1[t]` having a fixed value, it must be regenerated with this value cleared. This is done with the following functions:

```
pH1[t] =.;
pars = {{pH1[t], 7.2}};
p = {pH1[t]};
v̄ := VMatrix[eqns] ;
```

The concentration response matrix can now be calculated with

```
crm = ConcResponseMatrix[S̄, V̄N̄,
        v̄, p, pars, SteadyStateConc → steadyState];
```

And the response coefficient of 2,3-BPG concentration with respect to intracellular pH is

```
crm[[B23PG[t] /. substrateToNum]]
```

```
{47.304}
```

This number, in isolation, does not convey much information about the sensitivity of 2,3-BPG concentration to changes in pH. We need some means of comparing it with other values, and decomposing it into its component values; this is done by calculating partial response coefficients.

Q: Determine the contributions of the individual reactions in the erythrocyte model to the 2,3-BPG response to a change in pH.

A: To do this we use the **PartialConcResponse** function in the **Metabolic-ControlAnalysis** add-on package, as follows:

```
pcr = PartialConcResponse[S̄, V̄N̄, v̄, B23PG[t] /. substrateToNum,
        p, pars, SteadyStateConc → steadyState];
```

The following input then returns a list of the positions where the absolute value of the partial concentration response coefficient is ≥ 5.

```
bigPcrPosition =
  Position[pcr // Flatten, x_ ? (Abs[#1] ≥ 5 &)] // Flatten;
names = bigPcrPosition /. numToRxn ;
values = Part[pcr // Flatten, bigPcrPosition];
Transpose[{names, values}]
```

```
{{hk, 10.2622}, {pfk, 27.3347}, {bpgsp3, 7.86728}}
```

In conclusion, the enzymes most responsible for the response of 2,3-BPG concentration to pH are hexokinase, phosphofructokinase, and step 3 of the 2,3-BPG synthase-phosphatase reaction scheme.

Finally, the following function demonstrates that all the partitioned responses add up to the total value given in the previous example.

Apply[Plus, pcr]

```
{47.304}
```

This is a very reassuring result, not only with respect to our efficient use and understanding of MCA, but it is also a valuable check on the *Mathematica* programming of the various complicated functions.

Q: What is the value of the response coefficient of glycolytic flux with respect to intracellular pH?

A: The response of glycolytic flux to pH change is calculated by taking the flux via hexokinase as the relevant value. The analysis is as follows:

frm = FluxResponseMatrix[S̄, V̄N̄,
 v̄, p, pars, SteadyStateConc → steadyState,
 Normalized → zeroFluxPosition];

frm[[hk /. fluxToNum]]

```
{-1.89546}
```

Again, this value has little meaning unless it can be analyzed further to determine which reactions are the major contributors to it. This analysis is done by determining the partial flux response coefficients, as is shown in the following question/answer.

Q: Partition the total flux response coefficient given in the previous question/answer.

A: To address this question we must calculate the partial flux response coefficients; this is done as follows:

```
pfr = PartialFluxResponse[s̄, VN̄, v̄,
    hk /. rxnToNum, p, pars, SteadyStateConc → steadyState,
    Normalized → zeroFluxPosition] ;
```

A list in which the absolute value of the coefficient is ≥ 0.01 is generated as follows:

```
bigPcrPosition =
 Position[pfr // Flatten, x_ ? (Abs[#1] ≥ 0.01 &)] // Flatten;
names = bigPcrPosition /. numToRxn ;
values = Part[pfr // Flatten, bigPcrPosition] ;
Transpose[{names, values}]
```

```
{{hk, 0.0272152}, {pfk, 0.0724396}, {gapdh, -0.0438938},
 {pk, -0.048605}, {ldh, 0.0402698}, {bpgsp1, 0.230263},
 {bpgsp2, -0.515469}, {bpgsp3, -2.9455},
 {bpgsp4, -0.0179343}, {bpgsp7, -0.388286},
 {phostransport, -0.139795}, {mgatp, 0.185179},
 {mgadp, 0.127097}, {mgb23pg, -0.448401}, {hbmgatp, 1.82561},
 {hbatp, 1.07457}, {hbadp, 0.527308}, {hbbpg, -1.45574}}
```

Thus, there are 18 reactions through which intracellular pH exerts a significant effect on the glycolytic flux that occurs via hexokinase. Take some time to think about this finding and endeavor to form a metabolic overview of this control.

It is relevant to check, as was done for one of the flux control coefficients in Section 8.4, that the response coefficient calculated above agrees with that estimated by numerical modulation.

Q: Calculate the response coefficient of glycolytic flux with respect to intracellular pH by numerical modulation.

A: This is done as follows:

```
solution1 = NDSolveMatrix[s̄, VN̄, v̄, ic1, {t, 0, 1 × 10⁶},
    pars, AccuracyGoal → 10, PrecisionGoal → 10,
    WorkingPrecision → 15, MaxSteps → 2000] ;

v1 = v[hk] /. pH1[t] → 7.2 /. t → 1 × 10⁶ /. solution1;


pars2 = {{pH1[t], 1.001 * 7.2}};
(*Define new value for pH1[t].*)

solution2 = NDSolveMatrix[s̄, VN̄, v̄, ic1, {t, 0, 1 × 10⁶},
    pars2, AccuracyGoal → 10, PrecisionGoal → 10,
```

```
WorkingPrecision → 15, MaxSteps → 2000];
(*Determine the steady state after parameter modulation.*)

v2 = v[hk] /. pH1[t] → 7.2 1.001 /. t → 1 × 10⁶ /. solution2;
(*Determine the velocity of hk after parameter modulation.*)
```

$$\text{ResponseCoeff} = \frac{v2 - v1}{v1} \times \frac{1}{0.001}$$

```
{-1.93932}
```

This value is in very good agreement with that calculated above using the matrix method.

8.7 Elasticity Coefficients

We can also use the **PiElasticityMatrix** and the **EpsilonElasticity-Matrix** functions to calculate the whole matrices of elasticity coefficients (see Chapter 5). Note that with these functions **Normalized** can have all the usual Options (see Appendix 2 for details).

Q: Calculate the non-normalized π-elasticity matrix (Eqn [5.13]) with respect to intracellular pH that is used as a parameter here.

A: Simply apply the following function to the erythrocyte model; recall that the model is defined by the vectors and matrices $\bar{S}, \overline{VN}, \bar{v}, p, \text{pars}$.

```
PiElasticityMatrix[S̄, VN̄, v̄, p, pars,
SteadyStateConc → steadyState, Normalized → False]

{{1.788 × 10⁻⁷}, {0}, {5.97536 × 10⁻⁶}, {0}, {0}, {0.0000498037},
 {0}, {0}, {0}, {-9.37433 × 10⁻⁸}, {-0.0000292711},
 {0}, {-0.000014607}, {8.51056 × 10⁻⁷}, {0.0360437},
 {3.25555 × 10⁻⁸}, {0}, {0}, {0.00271634}, {0}, {0}, {0},
 {0}, {0}, {0}, {0}, {0}, {0}, {0}, {0}, {0}, {0}, {0}, {0},
 {-6.92731 × 10⁻⁶}, {0}, {0}, {0}, {8.8578 × 10⁻¹⁰}, {0},
 {8.07328 × 10⁻⁸}, {0.615534}, {0.0408956}, {0.657863},
 {0.0000225723}, {0}, {0}, {0.00710249}, {-0.978296},
 {-0.907653}, {-0.445397}, {-12.1412}, {-0.00186074}}
```

It is clear that a significant number of the reactions have no pH dependence. This is obvious from the form of the expressions that we specified in the first place. The reactions that are pH dependent, however, do not always have a pH dependence that is readily apparent, and this is due to the actual values of the concentrations of all the reactants at the steady state.

Q: Calculate all non-normalized ε-elasticity coefficients (Eqn [5.12]) for each reaction with respect to 2,3-BPG.

A: The **MetabolicControlAnalysis** add-on package has the relevant function **EpsilonElasticityMatrix** to automatically carry out this operation.

```
eem = EpsilonElasticityMatrix[S̄, V̄N̄, v̄, pars,
    SteadyStateConc → steadyState, Normalized → False];
```

```
Dimensions[eem]
```

```
{53, 56}
```

The latter two numbers are the dimensions of the matrix, **eem**, of ε-elasticity coefficients. Now pick out all the elasticities for each reaction with respect to 2,3-BPG from the matrix that are ≥ 0.0000001.

```
bigEcPosition = Position[eem[[All,B23PG[t]]/.substrateToNum]] // Flatten,
    x_ ? (Abs[#1] ≥ 0.0000001 &)] // Flatten;
names = bigEcPosition /. numToRxn ;
values =
    Part[eem[[All,B23PG[t]]/.substrateToNum]] // Flatten, bigEcPosition];
Transpose[{names, values}]
```

```
{{hk, -4.04733×10⁻⁶}, {pfk, -0.000261023},
 {ald, -0.0000645999}, {bpgsp7, -0.558427},
 {mgb23pg, 309.097}, {hbbpg, 1116.85}}
```

Thus, it is seen that there are six major 2,3-BPG sensitive steps.

8.8 Internal Response Coefficients

Partial internal response coefficients can also give valuable information on the control of a metabolic pathway. For example, the homeostatic strength of a reaction gives information about how important that reaction is in controlling the concentration of a particular metabolite (Section 5.6).

Q: Calculate the homeostatic strength of all reactions with respect to 2,3-BPG concentration.

A: The parameter that is sought is the partial internal response coefficient with respect to [2,3-BPG] for each reaction flux. We can calculate these as follows:

```
num = B23PG[t] /. substrateToNum;

picr = PartialInternalConcResponse[S̄, V̄N̄, v̄,
    num, num, pars, SteadyStateConc → steadyState];

bigPcIRPosition =
  Position[picr, x_ ? (Abs[#1] ≥ 0.01 &)] // Flatten;

names = bigPcIRPosition /. numToRxn;
values = Part[picr, bigPcIRPosition];
Transpose[{names, values}]
```

```
{{hk, -0.0952908}, {pfk, -0.489822}, {bpgsp7, -0.155465},
  {mgb23pg, -0.188363}, {hbbpg, -0.0707982}}
```

Thus, hexokinase, phosphofructokinase, and the seventh reaction of 2,3-BPG synthase-phosphatase are almost all equally important in controlling the concentration of 2,3-BPG. This analysis also reveals that the binding of 2,3-BPG to free magnesium and hemoglobin are very important for controlling the concentration of 2,3-BPG.

The accuracy of the calculations is confirmed by adding all homeostatic strengths; the correct result is −1 .

```
Apply[Plus, picr]

-1.
```

The partial response coefficients are the high-order metabolic control analysis parameters as they provide insights that would be hard to obtain by less formal or non-mathematical means.

8.9 Concluding Remarks

The odyssey, if you have read this book from its beginning to this endpoint, has taken you from the principle of mass action in chemical kinetics, through enzyme kinetics and simulation of simple enzymic reactions, to a contemporary model of one aspect of mammalian metabolism. It has culminated in the analytical dissection of the control of this model. We would have liked to have paused more to consider, on the way, additional questions and examples and develop greater familiarity with various aspects of MCA, in particular, but space and time did not allow this. We are, however, confident that we have presented all the key elements of a strategy for the simulation and metabolic control analysis of realistic metabolic systems; and in addition we have also pointed the way to an approach for parameter estimation that will be the key to linking metabolic simulation to the experiments which inform the next iteration in the refinement of a model.

8.10 Exercises

8.10.1

Comment on the significance of the fact that hexokinase, phosphofructokinase, and 2,3-BPG synthase-phosphatase step 3 are the main controllers of the normal steady-state concentration of 2,3-BPG.

8.10.2

Is the result in Section 8.6, that there are 18 reactions through which intracellular pH exerts a significant effect on glycolytic flux, surprising to you? Is it a result that other scientists and scholars might have expected?

8.10.3

Give a verbal description of the results obtained in the last example of Section 8.7. Give a biochemical explanation of the functional implications of the results for a red blood cell.

8.10.4

This is very open-ended (!): explore the effects on any metabolic control analysis parameter that arises as the consequence of the alteration of the activity of one or more enzymes, such as may occur with an inborn error of metabolism. Investigate if there might be metabolic strategies to circumvent the clinical consequences of the defect, based on insights gained from metabolic control analysis of the model.

8.11 References

1. Mulquiney, P.J. and Kuchel, P.W. (1997) Model of the pH-dependence of the concentrations of complexes involving metabolites, haemoglobin and magnesium ions in the human erythrocyte. *Eur. J. Biochem.* **245**, 71-83.
2. Mulquiney, P.J., Bubb, W.A., and Kuchel, P.W. (1999a) Model of 2,3-bisphosphoglycerate metabolism in the human erythrocyte based on detailed enzyme kinetic equations: *in vivo* kinetic characterization of 2,3-bisphosphoglycerate synthase/phosphatase using ^{13} C and ^{31} P NMR. *Biochem. J.* **342**, 567-580.
3. Mulquiney, P.J. and Kuchel, P.W. (1999b) Model of 2,3-bisphosphoglycerate metabolism in the human erythrocyte based on detailed enzyme kinetic equations:

computer simulation and metabolic control analysis. *Biochem. J.* **342**, 597-604.
4. Mulquiney, P.J. and Kuchel, P.W. (1999c) Model of 2,3-bisphosphoglycerate metabolism in the human erythrocyte based on detailed enzyme kinetic equations: equations and parameter refinement. *Biochem. J.* **342**, 581-596.
5. Heinrich, R. and Schuster, S. (1996) *The Regulation of Cellular Systems*. Chapman & Hall, New York.

Appendix 1 - Rate Equation Deriver

`RateEquation[rcm,el]` derives the steady-state rate equation for an enzyme mechanism defined in the rate constant matrix (rcm – see below). The argument el is optional and is a list of user-defined names for the enzyme forms of the reaction mechanism.

Rate constant matrix (rcm): this is constructed by drawing a square grid with n × n cells, where n is the total number of enzyme forms (free enzyme plus complexes) in the reaction scheme. On the left, adjacent to the first column, form a list down through each enzyme form, placing the names at consecutive rows with an arrow facing left to right after each name; this labels each "from" row. Then, across the top of the first row write the names of each enzyme form, in the same order as used for the first column. Before each enzyme name place an arrow facing left to right; this labels the "to" columns. Refer to the enzyme reaction scheme and place the relevant rate constant, together with any relevant reactant symbol, in the appropriate grid location. Note that a zero appears in all cells at the intersection of a "from" row with the "to" column of the same species. See also Section 3.4.2 for a description of this matrix.

The program is based on the method described by Cornish-Bowden.[1] The rationale of the method is summarized as follows: in the steady-state the proportion of any enzyme form, $E(m)$, can be written as $[E(m)]/e_0 = N(m)/D$ where e_0 is the total enzyme concentration, $N(m)$ is the specific numerator expression, and D equals the sum of all $N(m)$. These equations are known as the distribution equations.

The rules to determine $N(m)$ that can be inferred from the method of King and Altman[2] are
(1) Each $N(m)$ is a product of n-1 rate constants where n is the number of enzyme forms.
(2) There are no rate constants for reactions leading directly away from $E(m)$.
(3) There is one rate constant only for a reaction leading directly away from each enzyme form apart from $E(m)$ itself.
(4) There is at least one rate constant for a reaction leading to $E(m)$.
(5) No cyclic reactions can be represented in the product.

The following program generates products of rate constants that satisfy Rules 1 – 5 and hence determines the steady-state rate equation. In its present form the program can only determine rate equations for reaction mechanisms containing <15 enzyme forms.

```
RateEquation[rcm_, iel_ : {}] := Module[
    {el, rcm2, rcm3, sl, xx, yy, rate, bm, combm, nm2, nm3, p5, cpa,
```

```
        i, j, l, n, o, X, cpm, nmf, npm, en, denom1, denom2, num, dd},
(*Generate list of enzyme forms (el)
   if not entered as input.*)
If[iel == {}, el = Table[e[i], {i, Length[rcm]}], el = iel];
(* Subroutine to generate substrate list (sl). *)
(*Convert rcm to a list
   containing rate constants and substrates.*)

  rcm2 = Flatten[rcm /. Times → List];
(*Delete all rate constants in this new list.*)
  rcm3 = DeleteCases[rcm2, x_Subscript];

    rcm4 = DeleteCases[rcm3, x_k];
(*Delete all rate constants which are zero.*)
  sl = DeleteCases[rcm4, 0];
(*Subroutine to generate the rate equation in terms of
   concentrations of enzyme forms for each substrate.*)
  rm = Table[rcm[[i]] * el[[i]], {i, Length[rcm]}];
(*Determine the positions of the rates
   which contain substrates. These positions
     give the rates of the steps that
   consume the substrate.*)
  xx[1] = Table[Position[rm, sl[[i]] * b_], {i, Length[sl]}];
(*Determine the positions
   of the rates which release substrate.*)
  yy[1] = Table[Reverse[xx[1][[i, j]]],
  {i, Length[xx[1]]}, {j, Length[xx[1][[i]]]}];
(*Extract the rates of the steps which
   consume the substrate.*)
  xx[2] = Table[Extract[rm, xx[1][[i, j]]],
  {i, Length[xx[1]]}, {j, Length[xx[1][[i]]]}];
(*Extract the rates of the steps which
   produce the substrate.*)
  yy[2] = Table[Extract[rm, yy[1][[i, j]]],
  {i, Length[yy[1]]}, {j, Length[xx[1][[i]]]}];
(*Add all the rates together which
   consume a particular substrate.*)
  xx[3] = Table[xx[2][[i]] /. List → Plus,
  {i, Length[xx[2]]}];
(*Add all the rates together which
   produce a particular substrate.*)
  yy[3] = Table[yy[2][[i]] /. List → Plus,
  {i, Length[yy[2]]}];
(*Generate the rates by subtracting the
```

rates of steps that consume a particular
  substrate from those that produce it.*)
 Do[
 rate[sl[[i]]] = yy[3][[i]] - xx[3][[i]], {i, Length[sl]}];
(* Number of enzyme forms.*)
n = Length[rcm];
(* To create the bm table we first omit
  the "from E" (E→) ROW; this ensures that the
  ultimate product has no step leading AWAY from E;
 but it includes every other row thus ensuring
   that there is exactly one step leading
 away from every other enzyme form. The
   process is repeated for the other rows that
   correspond to the EA→, EAB->, etc species.*)
bm = Table[Delete[rcm, i], {i, n}];
(* Form a table of all possible list COMBINATIONS of
    rate constants at level 1 of bm... for this we
  define an outer product function that operates
  on lists (n-tuples) rather than products of
  rate constants. Define fn[a,b,c,...] where a,
 b,c, ... are the first, second,
 third ... n-tuples of each entry in bm. Thus,
 generate n-tuples of rate constants that satisfy Rules 1,
 2 and 3 for each enzyme form.*)
fn[a_, b_: {1}, c_: {1}, d_: {1}, e_: {1}, f_: {1}, g_: {1}, h_:
    {1}, i_: {1}, j_: {1}, k_: {1}, l_: {1}, m_: {1}, n_: {1}] :=
DeleteCases[Flatten[Outer[List, a, b, c, d, e,
    f, g, h, i, j, k, l, m, n], 13], 1, 2];
(*The function generates outer products for up to 14
  n-tuples each of 14 rate constants.
    The n-tuples are given the default value of {1} so that
    Flattening this list and deleting the 1's eliminates
    any extra entries in the 'default argument list'.*)
combm = Apply[fn, bm, {1}];
(*Eliminate n-tuples containing zeroes.This requires
  searching for an intersection of an n-tuple with
  {0}. If this is "Null" the n-tuple is retained.*)
nm2 = DeleteCases[Table[If[Intersection[combm[[i, j]], {0}] !=
    {0}, combm[[i, j]], {dd}], {i, Length[combm]},
    {j, Length[combm[[i]]]}], {dd}, Infinity];
(*Eliminate n-tuples that do not contain rate
  constants that lead to the respective
  enzyme forms in each 'row' of nm2.*)
p1 = Table[Position[nm2[[i]], rcm[[j, i]]], {i, n}, {j, n}];

```
p2 = Table[Flatten[p1[[i]], 1], {i, n}];
p3 = Table[p2[[j, i, 1]], {j, n}, {i, Length[p2[[j]]]}];
p4 = Table[Union[p3[[i]]], {i, n}];

nm3 = Table[nm2[[i, p4[[i, 1]]]], {i, n}, {1, Length[p4[[i]]]}];
```
(* Subroutine to create a matrix of n-tuples that
      satisfies Rules 1-5: i.e., remove cyclic
   n-tuples from nm3. This uses three nested loops, the
   outer one being to count through each enzyme form.*)
```
For[i = 1, i < n + 1, i++,  (*nm3 contains lists
```
   which are in turn lists of the product
   n-tuples for each enzyme form. This 'line' creates
   a loop to count from one enzyme form to the next. i.e.,
 to count through from nm3[[1]] to nm3[[n]].*)
```
 For[j = 1, j < Length[nm3[[i]]] + 1, j++,
```
  (*Loop to count through the n-
    tuples in the list for each enzyme form.*)
```
  For[m = 1, m < n, m++,  (* Loop to step through each
```
      element (rate constant) in each n-tuple in nm3.*)
  (* Return the column number (i.e., index of the *to*-
      enzyme form) in rcm that each element of each
      n-tuple is in. Note that Position[rcm,nm3[[i,j,m]]]
      gives the list {{row,column}} defining the
      position of the n-tuple element defined by
      nm3[[,j,,m]] in rcm. The suffix [[1,2]] to this
      command then picks out the column number.*)
```
  p5[i, j, m, 1] = Position[rcm, nm3[[i, j, m]]][[1, 2]];
```
  (* If the rate constant index, i,
    corresponds to the enzyme form index then the
    n-tuple is considered to contain a rate-constant
    that leads *to* the requisite enzyme form.*)
```
  If[p5[i, j, m, 1] == i, Continue[]];
```
  (*If the element of the product
      n-tuple is in column i of rcm then increment m.Or
      in other words if the element of the product n-
      tuple leads to the enzyme form in column i,
    then check the next element of the product.If this
    is not the case perform the next subroutine. Note
    that the command Continue tells the computer
    to increment the last loop. i.e.the k loop.*)
  (* The following subroutine determines whether the
      element of the product n-tuple (nm3[[i,j,m]]) leads
      to other elements of the product n-tuple which
      eventually lead to the enzyme form in column i.*)

```
(o = 1; Label[xxx]; X = o + 1;
 p6 = Flatten[Table[Position[
     rcm[[p5[i, j, m, o]]], nm3[[i, j, t]]], {t, n - 1}]];
```
(* We previously determined the enzyme form that the element
   of the product n-tuple (nm3[[i,j,m]]) led to. This
   next step determines which rate constant of the
   n-tuple leads away from this enzyme form and
   hence what enzyme form this rate constant
   leads to. This is achieved in the following
   way. For example if nm3[[i,j,m]] leads to
   the enzyme form in *column* y,we then determine
   which column contains an element of the n-
   tuple in *row* y. Thus using the Table command
   we jump through each element of the n-tuple
   (as defined by nm3[[i,j,t]] in the above command)
   and find its position in the row defined by
   rcm[[p5[i,j,m,o]]].*)
```
If[p6 === {}, cpa[i, j] = nm3[[i, j]]; m = n; Continue[]];
```
(* If the enzyme form in the column which contains
   element nm3[[i,j,m]] leads to nothing i.e.,
   if P6=={} then the product is cyclic. We then
     set the array element cpa[i,j] equal to
     this cyclic product n-tuple. m is set to n,
   so that the k loop for these particular values
   of i and j finishes. The continue command
   then increments to the next value of j.*)
```
 p5[i, j, m, X] = p6[[1]];
 If[p5[i, j, m, X] == i, Continue[]];
```
(*If the enzyme form in the column which contains
   element nm3[[i,j,m]] leads to the enzyme form i then
   increment m (do this via the Continue command). If
   not keep following the product through
   until enzyme form i is found. If it is
   not found then the product is cyclic.*)
```
 o++;
 If[o < n, Goto[xxx]]);
 cpa[i, j] = nm3[[i, j]];
];

];

];

(* Turn cpa (cyclic product array) into cpm
```

```
(cyclic product matrix). The cpa[x_,y_] part of the
    composite function is there to specify elimination
    of the cyclic terms in the formation of CPM.*)
  cpm = DeleteCases[Table[cpa[i, j], {i, n},
      {j, Length[nm3[[i]]]}], cpa[x_, y_],
Infinity];
  (*Remove cyclic n-
    tuples from nm3 to give the numerator product
matrix which satisfies Rules 1-5.*)
  nmf = Table[Complement[nm3[[i]], cpm[[i]]], {i, n}];
  npm = Table[Apply[Times, nmf[[i, j]]],
      {i, n}, {j, Length[nmf[[i]]]}];
  Do[en[i] = Apply[Plus, npm[[i]]], {i, n}];
  denom1 = Apply[Plus, Flatten[npm]];
  denom2 = Collect[denom1, s1];
  num = Simplify[Table[rate[s1[[i]]] , {i, Length[s1]}] /.
      Table[el[[i]] → en[i], {i, Length[el]}]] ;

  Print["Enzyme Distribution Functions"];
  Do[Print[el[[i]], "/eo = " en[i]], {i, n}];
  Print[""];
  Print["Steady-State Rate Equations"];
    Do[Print["d"[s1[[i]]], "/dt =",
    eo num[[i]], "/ Denominator"], {i, Length[s1]}];
  Print[""];
  Print["Denominator"];
  Print[denom2];
  ];


Save["rateequationderiver", RateEquation];
```

References

1. Cornish-Bowden, A. (1977) An automatic method for deriving steady state rate equations. *Biochem. J.* **165**, 55-59.
2. King, E.L. and Altman, C. (1956) A schematic method of deriving the rate laws for enzyme catalyzed reactions. *J. Phys. Chem.* **60**, 1375-1378.

Appendix 2 - Metabolic Control Analysis Functions

A2.1 Metabolic Control Analysis Functions

The following is an alphabetical listing of the 21 functions included in the add-on package **MetabolicControlAnalysis**. These functions can be used after applying the command, **<<MetabolicControlAnalysis`**. This listing of functions is followed by the program used to create the package **MetabolicControlAnalysis**.

ConcControlMatrix

ConcControlMatrix[S, N, v, p, SteadyState→*steadystate*] calculates a matrix for the metabolic system defined by **S, N, v,** and **p,** at the steady state given by the replacement rule *steadystate,* where the element m_{ij} is the normalized concentration control coefficient of metabolite i with respect to reaction j.

- Inclusion of the parameter table **p** is optional.
- *steadystate* in the argument SteadyStateConc→*steadystate* must have the form of a replacement rule as generated by **SteadyState** or **NSteadyState**.
- The default value for the Option, SteadyStateConc, is **S**.
- The Option Normalized→False can be included to calculate non-normalized control coefficients.
- See also: **FluxControlMatrix**.

ConcResponseMatrix

- **ConcResponseMatrix[S, N, v,** {*parameter list*}, **p,** SteadyState→*steadystate*] returns a matrix where the element m_{ik} is the concentration response coefficient of the concentration of metabolite i in **S** with respect to the kth parameter of {*parameter list*}.

- Inclusion of the parameter table **p** is optional.
- *steadystate* in the argument SteadyStateConc→*steadystate* must have the form of a

replacement rule as generated by **SteadyState** or **NSteadyState**.
- The default value for the Option, SteadyStateConc, is **S**.
- The Option Normalized→False can be included to calculate non-normalized response coefficients.
- See also: **FluxResponseMatrix**.

ConservationRelations

- **ConservationRelations[S, N]** determines non-negative conservation relations between the metabolites S_i in the metabolic network defined by **N**.

- The Option, GMatrix→True, will return a matrix, **G**, such that **G.S = Const** where **Const** is a matrix of constants. The default value for GMatrix is False.
- All entries in **N** must be exact numbers.
- See also: **NSteadyState**.

EpsilonElasticityMatrix

- **EpsilonElasticityMatrix[S, N, v, p,** SteadyState→*steadystate*] returns a matrix where the element m_{ij} is the ε-elasticity of reaction i in **v** with respect to substrate j in **S**.

- Inclusion of the parameter table **p** is optional.
- *steadystate* in the argument SteadyStateConc→*steadystate,* must have the form of a replacement rule as generated by **SteadyState** or **NSteadyState**.
- The default value for the Option, SteadyStateConc, is **S**.
- The Option Normalized→False can be included to calculate non-normalized elasticity coefficients.
- The Option Normalized→zerofluxposition, where zerofluxposition is a list $\{\{i\},\{j\},$ ....\}, can be included so that normalized coefficients are *not* calculated for fluxes through reactions $i, j, \ldots$. This is important when there are zero fluxes in the system as the normalized coefficients become undefined.
- See also: **PiElasticityMatrix**.

FluxControlMatrix

- **FluxControlMatrix[S, N, v, p,** SteadyState→*steadystate*] calculates a matrix for the metabolic system defined by **S, N, v,** and **p,** at the steady state given by the replacement rule *steadystate,* where the element m_{ij} is the normalized flux control coefficient of the flux through reaction i with respect to reaction j.

- Inclusion of the parameter table **p** is optional.
- *steadystate* in the argument SteadyStateConc→*steadystate,* must have the form of a

replacement rule as generated by **SteadyState** or **NSteadyState**.
- The default value for the Option, SteadyStateConc, is **S**.
- The Option Normalized→False can be included to calculate non-normalized control coefficients.
- The Option Normalized→zerofluxposition, where zerofluxposition is a list $\{\{i\},\{j\}, ...\}$, can be included to so that normalized coefficients are *not* calculated for fluxes through reactions $i, j, ...$. This is important when there are zero fluxes in the system as the normalized coefficients become undefined.
- See also: **ConcControlMatrix**.

FluxResponseMatrix

- **FluxResponseMatrix[S, N, v,** *{parameter list}*, **p**, SteadyState→*steadystate*] returns a matrix where the element m_{jk} is the flux response coefficient of the flux through reaction j in **v** with respect to the kth parameter of *{parameter list}*.

- Inclusion of the parameter table **p** is optional.
- *steadystate* in the argument SteadyStateConc→*steadystate,* must have the form of a replacement rule as generated by **SteadyState** or **NSteadyState**.
- The default value for the Option, SteadyStateConc, is **S**.
- The Option Normalized→False can be included to calculate non-normalized response coefficients.
- The Option Normalized→zerofluxposition, where zerofluxposition is a list $\{\{i\},\{j\}, ...\}$, can be included so that normalized coefficients are *not* calculated for fluxes through reactions $i, j, ...$ This is important when there are zero fluxes in the system as the normalized coefficients become undefined.
- See also: **ConcResponseMatrix**.

LinkMatrix

- **LinkMatrix[S, N]** rearranges the rows of **N** so that its upper rank(**N**) rows are linearly independent and form a submatrix **N**°. Output of this function is a table. The first element of the table gives the new **S**. The second gives a 'Link Matrix', **L**, such that $\mathbf{N} = \mathbf{L}\,\mathbf{N}° = \begin{pmatrix} \mathbf{I} \\ \mathbf{L} \end{pmatrix}\mathbf{N}°$. The third row gives the **N**° matrix, the fourth the rearranged **N**. The fifth element contains the matrix **G** which has the property **GN = 0**, and the last row gives the transformation rules for transforming the old matrices into the new matrices. Each transformation rule is in the form {old row number, new row number}.

- See Section 4.9 for more information.

MMatrix

- **MMatrix [S, N, v, p,** Normalized→False, SteadyStateConc→*steadystate*] calculates the Jacobian of the differential equation system defined by **S, N, v,** and **p.**

- Note that the last three arguments are optional, however, the default value for Normalized is True.
- See also: **Stability**.

NDSolveMatrix

- **NDSolveMatrix [S, N, v,** *initial conditions,* {*t,tmin,tmax*}] uses the function **NDSolve** to find a numerical solution for the metabolite concentrations, **S,** with time in the range *tmin* to *tmax,* for a system of ordinary differential equations defined by the matrices **S, N,** and **v,** and subject to the *initial conditions.*
- **NDSolveMatrix [S, N, v,** *initial conditions,* {*t,tmin,tmax*}, **p**] solves the system of ordinary differential equations defined by the matrices **S, N, v,** and the parameter matrix **p. p** has the form {{p_1, *value*$_1$ }, {p_2, *value*$_2$ }, ...}.

- **NDSolveMatrix** has the same Options as **NDSolve**.

NMatrix

- **NMatrix [eqn, extpars]** generates a numerical-only stoichiometry matrix for the reaction system defined in the equation list, **eqn**; it takes into account the parameters in the list, **extpars**, are external parameters.

- See also: StoichiometryMatrix.

NSteadyState

- **NSteadyState [S, N, v, p, init]** uses **FindRoot** to determine an approximate numerical solution to **N v** = **0** for a system of ordinary differential equations defined by the matrices **S, N, v,** and **p. init** contains initial estimates of the steady-state concentrations in the form of a replacement rule.

- Inclusion of the parameter table **p** is optional.
- **NSteadyState** gives solutions in terms of rules of the form $x \rightarrow sol$.
- The constants returned by ConservationRelations must be assigned values before **NSteadyState** can be applied.
- **NDSteadyState** has the same Options as **FindRoot**.
- See also: **NSteadyState**, **NDSolveMatrix**, **ConservationRelations**.

PartialConcResponse

- **PartialConcResponse[S, N, v,** n, *{parameter list}*, **p**, SteadyState $\rightarrow$ *steadystate*] returns a matrix of partial concentration response coefficients for a metabolite at position n in **S**. Each entry m_{jk} in the matrix gives the product of the concentration control coefficient with respect to reaction j and the π-elasticity coefficient with respect to reaction j and parameter k.

- Inclusion of the parameter table **p** is optional.
- *steadystate* in the argument SteadyStateConc→*steadystate*, must have the form of a replacement rule as generated by SteadyState or NSteadyState.
- The default value for the Option, SteadyStateConc, is **S**.
- The Option Normalized→False can be included to calculate non-normalized partial response coefficients.
- See also: **PartialFluxResponse**.

PartialFluxResponse

- **PartialFluxResponse[S, N, v,** n, *{parameter list}*, **p**, SteadyState $\rightarrow$ *steadystate*] returns a matrix of partial flux response coefficients for a flux at position n in **v**. Each entry m_{jk} in the matrix gives the product of the flux control coefficient with respect to reaction j and the π-elasticity coefficient with respect to reaction j and parameter k.

- Inclusion of the parameter table **p** is optional.
- *steadystate* in the argument SteadyStateConc→*steadystate*, must have the form of a replacement rule as generated by SteadyState or NSteadyState.
- The default value for the Option, SteadyStateConc, is **S**.
- The Option Normalized→False can be included to calculate non-normalized partial response coefficients.
- The Option Normalized→zerofluxposition, where zerofluxposition is a list $\{\{i\},\{j\}, ...\}$, can be included so that normalized coefficients are *not* calculated for fluxes through reactions $i, j, \ldots$. This is important when there are zero fluxes in the system as the normalized coefficients become undefined.
- See also: **PartialConcResponse**.

PartialInternalConcResponse

■ **PartialInternalConcResponse[S, N, v,** *n, m,* **p,** SteadyState → *steadystate***]** returns a vector which contains the partial internal concentration response coefficients for a metabolite at position *n* in **S** with respect to a metabolite at position *m* in **S**. The *j*th position in the vector is the partial internal response coefficient for reaction *j*.

■ Inclusion of the parameter table **p** is optional.
■ *steadystate* in the argument SteadyStateConc→*steadystate,* must have the form of a replacement rule as generated by **SteadyState** or **NSteadyState**.
■ The default value for the Option, SteadyStateConc, is **S**.
■ The Option Normalized→False can be included to calculate non-normalized partial internal response coefficients.
■ See also: **PartialInternalFluxResponse**.

PartialInternalFluxResponse

■ **PartialInternalFluxResponse[S, N, v,** *n, m,* **p,** SteadyState → *steadystate***]** returns a vector which contains the partial internal flux response coefficients for a flux at position *n* in **v** with respect to a metabolite at position *m* in **S**. The *j*th position in the vector is the partial internal response coefficient for reaction *j*.

■ Inclusion of the parameter table **p** is optional.
■ *steadystate* in the argument SteadyStateConc→*steadystate,* must have the form of a replacement rule as generated by **SteadyState** or **NSteadyState**.
■ The default value for the Option, SteadyStateConc, is **S**.
■ The Option Normalized→False can be included to calculate non-normalized partial internal response coefficients.
■ The Option Normalized→zerofluxposition, where zerofluxposition is a list $\{\{i\},\{j\},$ ....$\}$, can be included so that normalized coefficients are *not* calculated for fluxes through reactions *i, j,* ... . This is important when there are zero fluxes in the system as the normalized coefficients become undefined.
■ See also: **PartialInternalConcResponse**.

PiElasticityMatrix

■ **PiElasticityMatrix[S, N, v,** *{parameter list},* **p,** SteadyState → *steadystate***]** returns a matrix where the element m_{ik} is the π-elasticity of reaction *i* in **v** with respect to the *k*th parameter of *{parameter list}*.

■ Inclusion of the parameter table **p** is optional.
■ *steadystate* in the argument SteadyStateConc→*steadystate,* must have the form of a replacement rule as generated by **SteadyState** or **NSteadyState**.

■ The default value for the Option, SteadyStateConc, is **S**.

■ The Option Normalized→False can be included to calculate non-normalized elasticity coefficients.

■ The Option Normalized→zerofluxposition, where zerofluxposition is a list $\{\{i\},\{j\},$ ....\}, can be included so that normalized coefficients are *not* calculated for fluxes through reactions $i, j, ...$. This is important when there are zero fluxes in the system as the normalized coefficients become undefined.

■ See also: **EpsilonElasticityMatrix**.

SMatrix

■ **SMatrix[eqn, extpars]** generates the corresponding substrate list, **S**, for the reaction system defined by **eqn** and **extpars**.

■ See also: **NMatrix**, **VMatrix**.

Stability

■ **Stability[S, N, v, p**, SteadyStateConc → *steadystate*] assesses whether the differential equation system defined by, **N**, **v**, and **p** is asymptotically stable at the steady state given by the replacement rule *steadystate*. This function also returns the eigenvalues of the Jacobian of the differential equation system.

■ Inclusion of the parameter table **p** is optional.

■ *steadystate* in the argument SteadyStateConc→*steadystate,* must have the form of a replacement rule as generated by **SteadyState** or **NSteadyState**.

■ See also: **MMatrix**.

SteadyState

■ **SteadyState[S, N, v, p]** uses **Solve** to determine the solution to **N v** = 0 for a system of ordinary differential equations defined by the matrices **S, N**, **v**, and **p**.

■ Inclusion of the parameter table **p** is optional.

■ **SteadyState** gives solutions in terms of rules of the form x -> *sol*.

■ See also: **NSteadyState**, **NDSolveMatrix**.

StoichiometryMatrix

■ StoichiometryMatrix[**eqn, extpars**] is similar to NMatrix except that it returns a stiochiometry matrix which has rows and columns labelled by metabolite names and reaction names, respectively.

- See also: **NMatrix**.

VMatrix

- **VMatrix[eqn, extpars]** generates the corresponding reaction velocity list, **v**, for the reaction system defined by **eqn**.

- See also: **NMatrix**, **SMatrix**.

A2.2 Metabolic Control Analysis Program

The following is the program used to create the package **MetabolicControlAnalysis**.

```
Off[General::spell1];
Off[General::spell];


(*Begin MetabolicControl Analysis Package.*)
(*Note that Eqn numbers in the comments
   section refer to Eqns in Heinrich and Schuster,
   (1996) The Regulation of Cellular Systems.*)
BeginPackage["MetabolicControlAnalysis`",
   "LinearAlgebra`MatrixManipulation`"];


(*Usage messages for the exported functions.*)
MetabolicControlAnalysis::usage =
   "MetabolicControlAnalysis is a package of
     functions useful for metabolic control analysis.";

ConcControlMatrix::usage = "ConcControlMatrix[
     smatrix,nmatrix,vmatrix,pmatrix, options...]
   calculates a matrix of concentration control coefficients
     where matrix[[i,j]] is the concentration control
     coefficient of metabolite i with respect to reaction j.";

ConcResponseMatrix::usage =
   "ConcResponseMatrix[smatrix,nmatrix,vmatrix,{parameter
     list},pmatrix,options...] generates a matrix of
     concentration response coefficients where matrix[[
     i,j]] is the response coefficient of concentration
     i with respect to  kinetic parameter j.";

EpsilonElasticityMatrix::usage =
```

"EpsilonElasticityMatrix[smatrix,nmatrix,vmatrix,
   pmatrix,options...] generates a matrix where matrix[[
   i,j]] is the partial derivative of reaction i with
   respect to the concentration of substrate j.";

FluxControlMatrix::usage = "FluxControlMatrix[
   smatrix,nmatrix,vmatrix,pmatrix, options...]
 calculates a matrix of flux control coefficients
   where matrix[[i,j]] is the flux control
   coefficient of flux i with respect to reaction j.";

FluxResponseMatrix::usage =
  "FluxResponseMatrix[smatrix,nmatrix,vmatrix,
   {parameter list},pmatrix, options...] generates
   a matrix of flux response coefficients where
   matrix[[i,j]] is the response coefficient of
   flux i with respect to kinetic parameter j.";

LinkMatrix::usage =
  "LinkMatrix[smatrix,nmatrix] rearranges the rows of
   smatrix and nmatrix so that the upper rows of nmatrix
   are linearly independent (These linearly independent
   rows are called the submatrix N^0). Output of this
   function is a table. The first element of the table
   gives the new smatrix. The second gives the link matrix (
   L). The link matrix is a matrix which when right
   multiplied with N^0 gives the rearranged nmatrix. The
   third row gives the N^0 matrix and the last row give the
   transformation rules for Tranferring the old matrices
   into the new matrices. Each transformation rule in
   given in the form {old row number, new row number}.";

MMatrix::usage =
  "MMatrix[smatrix,nmatrix,vmatrix,pmatrix,options...]
   generates the Jacobian of the differential
   equation system described by smatrix,nmatrix,
   and vmatrix which has been approximated
   with a first order Taylor expansion.";

NDSolveMatrix::usage =
  "NDSolveMatrix[smatrix,nmatrix,vmatrix,initialconditions,
   {t,tmin,tmax},pmatrix,options...]finds a
   numerical solution to the ordinary differential
   equations described by smatrix, nmatrix, and

vmatrix for all metabolite in smatrix with the
independent variable t in the range tmin to tmax.";

NMatrix::usage =
"NMatrix[eqn,extpars] returns the stoichiometry matrix, or
    nmatrix, for the reaction system defined in the equation
    list, eqn, and taking into account the parameters
    in the list, extpars, are external parameters.";

ConservationRelations::usage =
    "ConservationRelations[smatrix,nmatrix]
    generates a list of conservation relations.";

NSteadyState::usage =
"SteadyState[smatrix,nmatrix,vmatrix,pmatrix] finds
    the steady state solution(s) of the system of
    ordinary differential equations described by smatrix,
    nmatrix,and vmatrix. A steady state implies that
    the concentrations of metabolites are constant.";

PartialConcResponse::usage =
    "PartialConcResponse[smatrix,nmatrix,vmatrix,
    n_Integer,{par},pmatrix,options...] finds
    a column vector of the partial response
    coefficients for metabolite n and parameter par.";

PartialFluxResponse::usage =
    "PartialFluxResponse[smatrix,nmatrix,vmatrix,
    n_Integer,{par},pmatrix,options...] finds
    a column vector of the partial response
    coefficients for flux n and parameter par.";

PartialInternalConcResponse::usage =
    "PartialInternalConcResponse[smatrix,nmatrix,vmatrix,
    n_Integer,m_Integer,pmatrix,options...] finds a column
    vector of the partial internal response coefficients
    for metabolite n with respect to metabolite m.";

PartialInternalFluxResponse::usage =
    "PartialInternalConcResponse[smatrix,nmatrix,vmatrix,
    n_Integer,m_Integer,pmatrix,options...] finds a
    column vector of the partial internal response
    coefficients for flux n with respect to metabolite m.";

```
PiElasticityMatrix::usage =
  "PiElasticityMatrix[smatrix,nmatrix,vmatrix,{paramter
    list},pmatrix, options...] generates a matrix
    where matrix[[i,j]] is the partial derivative of
    reaction i with respect to  kinetic parameter j.";

SMatrix::usage =
 "SMatrix[eqn, extpars] returns the substrate matrix
    for the reaction system defined in the equation
    list, eqn, and taking into account the parameters
    in the list, extpars, are external parameters.";

Stability::usage =
 "Stability[smatrix,nmatrix,vmatrix,pmatrix]  determines
    whether the steady state solution(s) of the system
    of  ordinary differential equations described by
    smatrix, nmatrix, and vmatrix is stable or unstable.";

SteadyState::usage =
 "SteadyState[smatrix,nmatrix,vmatrix,pmatrix] finds the
    steady state solution(s) of the system of  ordinary
    differential equations described by smatrix,
    nmatrix,and vmatrix. A steady state implies that
    the concentrations of metabolites are constant.";

StoichiometryMatrix::usage =
 "StoichiometryMatrix[eqn,extpars]
    returns an anotated stoichiometry matrix for the
    reaction system defined in the equation list,
    eqn, and taking into account the parameters in
    the list, extpars, are external parameters.";

VMatrix::usage =
 "VMatrix[eqn,extpars] returns a reaction velocity matrix for
    the reaction system defined in the equation list.";

(*Set default values of Options.*)
Options[EpsilonElasticityMatrix] =
  Options[MMatrix] =
   Options[PiElasticityMatrix] =
    Options[ConcControlMatrix] =
     Options[FluxControlMatrix] =
      Options[ConcResponseMatrix] =
       Options[FluxResponseMatrix] =
```

```
        Options[PartialConcResponse] =
         Options[PartialFluxResponse] =
          Options[PartialInternalConcResponse] =
           Options[PartialInternalFluxResponse] =
             {Normalized -> True, SteadyStateConc -> {}};

Options[ConservationRelations] =
  Options[ConservationRelations2] = {GMatrix -> False};

(*Usage messages for Options.*)
GMatrix::usage =
  "GMatrix is an option of ConservationRelations.
   The default value is false which returns the
   conservation relations. If this option is set
   to true the function returns the G matrix.";


Normalized::usage =
  "Normalized is an option of EpsilonElasticityMatrix,
   MMatrix,PiElasticityMatrix, ConcControlMatrix,
   FluxControlMatrix, ConcResponseMatrix,
   FluxResponseMatrix, PartialConcResponse,
   PartialFluxResponse,PartialInternalConcResponse,
   and PartialInternalFluxResponse.";


SteadyStateConc::usage =
  "SteadyStateConc is an option of EpsilonElasticityMatrix,
   PiElasticityMatrix, ConcControlMatrix,
   FluxControlMatrix, ConcResponseMatrix,
   FluxResponseMatrix, PartialConcResponse,
   PartialFluxResponse,PartialInternalConcResponse,
   and PartialInternalFluxResponse. Set to a list of
   replacement rules that define the steady state. The
   default value is S which returns the results of the
   function in terms of the steady state concentrations.";

(* Begin the private context.*)
Begin["Private`"] ;
(* Define auxiliary functions.*)
(*These functions are hidden from the package user
  and are only defined for programming purposes.*)

(*ConservationRelations2.*)
(*Determines a set of conservation relations using
  the G matrix calculated with LinkMatrix. Note that
```

the conservation constants, Konst, determined with
this function are different from those constants,
Const, determined with ConservationRelations.*)

```
ConservationRelations2[smatrix_List,
  nmatrix_List, opts___Rule] :=

Module[
  {transform, linkmatrixsolution, newgmatrix, gmatrixrule},

  (*Rearrange S, N, v into Eqn 3.8 using LinkMatrix.*)
  linkmatrixsolution = LinkMatrix[smatrix, nmatrix];
  newgmatrix = linkmatrixsolution[[5, 2]];
  (*The G matrix for the rearranged system.*)
  transform = linkmatrixsolution[[6, 2]];

  (*Assign True or False to gmatrixrule
    depending of the value of the GMatrix Option.*)
  gmatrixrule = GMatrix /. {opts} /.
    Options[ConservationRelations];

  (*Returns G or conservation relations,
   depending on value of gmatrixrule. Note that a gmatrix
    for the original system is returned. This involves
    rearranging the G matrix for the rearranged system
    using the function NewToOld described below.*)
If[newgmatrix == {},
    (*If gmatrix ={}, print {}*)
    {},
    (*Otherwise*)
    If[gmatrixrule,
      (*If gmatrixrule = True return gmatrix.*)
      Transpose[NewToOld[Transpose[newgmatrix], transform]],
      (*If gmatrixrule =
        False return conservation relations.*)
      Print[Transpose[NewToOld[Transpose[newgmatrix],
          transform]].smatrix, " == ",
       Array[Global`Konst, Length[newgmatrix]]]
        (*See Eqn 3.4 for this relation.*)
    ]
  ]
]
```

(*ConservationRelationsTransform.*)
(*Produces a replacement rule

which expresses the conservation constants,
Konst, determined with ConservationRelations2,
in terms of the nonnegative conservation constants,
Const, determined with ConservationRelations.*)
ConservationRelationsTransform[smatrix_List, nmatrix_List] :=

Module[{cr2, cr2i, rcr2i, transcr2, cr,
   cri, rcri, transcr, rkonst, ckonst, solution },

  cr2 =
   ConservationRelations2[smatrix, nmatrix, GMatrix -> True] ;

  If[cr2 == {}, {},
   (*If there are no conservation relations,
    the conservation relation transform rule
    is an empty set. If not determine the
    nonnegative conservation relations G matrix.*)
   cr = ConservationRelations[smatrix,
     nmatrix, GMatrix -> True] ;

   (*We now find two matrices, transcr and transcr2,
    such that transcr.cr == transcr2.cr2*)
   (*Note that RowReduce[cr]==
     RowReduce[cr2] and hence transcr and transcr2 can be
      determined by taking note of the row rearrangements
       that RowReduce performs on each of these matrices.*)

   (*determining transcr2.*)
   cr2i =
    Table[Join[cr2[[i]], IdentityMatrix[Length[cr2]][[i]]],
     {i, Length[cr2]}] (*Append identity matrix to cr2.*);
   rcr2i = RowReduce[cr2i] (*Rearrange cr2i with the function
     row reduce. The appended identity matrix in cr2i will
      now have been rearranged and will contain information
       about all the row combinations that RowReduce
        performed on cr2. Hence transcr2 is given by.*);
   transcr2 = Table[Take[rcr2i[[i]], -Length[cr2]],
     {i, Length[cr2]}];

   (*determining transcr.*)
   cri = Table[Join[cr[[i]], IdentityMatrix[Length[cr]][[i]]],
     {i, Length[cr]}] (*Append identity matrix to cr.*);
   rcri = RowReduce[cri] (*Rearrange cri with the
     function RowReduce. Hence transcr is given by.*);

```
  transcr = Table[Take[rcri[[i]], -Length[cr]],
    {i, Length[cr]}];

 (*Now since transcr.cr == transcr2.cr2, then
     transcr.const==transcr2.konst,
   where const is the vector of conservation constants
    associated with cr and konst is the vector of
    conservation constants associated with cr2.*)
 (*Hence we can determine the konsts in
    terms of the consts using Solve.*)
  rkonst = transcr2.Array[Global`Konst, Length[cr2]];
  cconst = transcr.Array[Global`Const, Length[cr]] // Flatten;

  solution = Solve[rkonst == cconst,
     Array[Global`Konst, Length[cr2]]][[1]]
 ]
]
```

```
(*LinkMatrixTransform.*)
(*Rearranges the rows of nmatrix so that the upper rows of
  nmatrix are linearly independent and then returns a
  table of transformation rules for Tranferring the old
  matrices into the new matrices. Each transformation rule
  is given in the form {old row number, new row number}.*)

LinkMatrixTransform[nmatrix_List] :=

Module[{tempmatrix, nomatrix, gmatrix, swap, numldrows},

  nomatrix = nmatrix;

  (*Determine number of linearly dependent rows.*)
  gmatrix = NullSpace[Transpose[nmatrix]];
  numldrows = Length[gmatrix];

  Do[ (*We need to find one transformation
     rule for each linearly dependent row.*)
    Do[
       (*Delete the ith row of nomatrix*)
       tempmatrix = Delete[nomatrix, i];
         (*Test to see whether
       the number of linearly dependent rows is
       decreased when the ith row is deleted.*)
       If[ Length[NullSpace[Transpose[tempmatrix]]] <
```

```
                Length[NullSpace[Transpose[nomatrix]]],
                (*If it is then row i is
            linearly dependent. Create swap rule ordered
            pair in the form {old row, new row}.*)
                swap[j] = {i + j - 1,
                        (Length[nmatrix] - numldrows + j)};
        (*Then break from this do statement to find the next
            transformation rule using the remaining rows*)
                nomatrix = tempmatrix;
                Break[]
            ];
        (*If row i is not linearly dependent then check i+1*)
        , {i, 2, Length[nomatrix]}];
    , {j, numldrows}];

    (*Create table of transformation rules.*)
    Table[swap[i], {i, numldrows}]
    ]


(*NewToOld.*)
(*Transforms a new matrix (N or S) into
  an old matrix using the transformation rules
  determined by LinkMatrixTransform.*)
NewToOld[matrix_List, transformmatrix_List] :=

Module[{deletetable, imatrix},

    (*Delete all the linearly dependent
      rows in the old matrix using transformmatrix*)
    deletetable = Table[{transformmatrix[[i, 2]]},
        {i, Length[transformmatrix], 1, -1}];
    imatrix = Delete[matrix, deletetable];

    (*Use TransformMatrix to reinsert the linearly
      dependent rows in their original positions.*)
    Do[imatrix = Insert[imatrix,
        matrix [[ transformmatrix[[i, 2]]]],
        transformmatrix[[i, 1]]]
      , {i, Length[transformmatrix]}];
    imatrix
]


(*OldToNew.*)
(*Transforms an old matrix (N or S) to
```

```
     a new matrix using the transformation rules
     determined by LinkMatrixTransform.*)
OldToNew[matrix_List, transformmatrix_List] :=

Module[{deletetable, imatrix, dmatrix, newmatrix},

  (*Delete all the linearly dependent
    rows in the matrix using transformmatrix*)
  deletetable =
   Table[{transformmatrix[[i, 1]]},
    {i, Length[transformmatrix]}];
  imatrix = Delete[matrix, deletetable];

  (*Create a table of linearly
    dependent rows using transform matrix.*)
  dmatrix = Table[matrix[[transformmatrix[[i, 1]]]],
    {i, Length[transformmatrix]}];

  (*Append dependent rows to independent rows.*)
  newmatrix = Flatten[{imatrix, dmatrix}, 1]
]

(*Define the exported functions.*)
(*Error messages for the exported functions *)
MetabolicControlAnalysis::badarg =
  "The number of `1` does not equal the
    number of `2` in the stoichiometric matrix!";

(*ConcControlMatrix.*)
ConcControlMatrix[smatrix_List, nmatrix_List,
  vmatrix_List, p_List: {}, opts___Rule] :=

 Module[{linkmatrixsolution, newsmatrix, linkmatrix,
    nomatrix, linkmatrixtransform, parameterrule,
    pars, parvals, steadystaterule, normalized,
    mmatrixar, coefmatrix, newsmatrixar, vmatrixar},

  (*Rearrange S, N, v into Eqn 3.8
    using LinkMatrix.*)
  linkmatrixsolution = LinkMatrix[smatrix, nmatrix];
  newsmatrix = linkmatrixsolution[[1, 2]];
  linkmatrix = linkmatrixsolution[[2, 2]];
  nomatrix = linkmatrixsolution[[3, 2]];
  linkmatrixtransform = linkmatrixsolution[[6, 2]];
```

```
(*Convert parameter list into a
   replacement rule called parameter rule.*)
 If[p == {},
parameterrule = {},
pars = Transpose[p][[1]]; parvals = Transpose[p][[2]];
parameterrule =
   Table[pars[[i]] -> parvals[[i]], {i, Length[pars]}]
];
```

```
(*Create the steady state replacement rule.*)
 steadystaterule =
 SteadyStateConc /. {opts} /. Options[ConcControlMatrix];
```

```
(*Calculate the nonnormalized concentration
   control matrix using Eqn 5.25 b and Eqn 5.24.*)
(*Eqn 5.24.*)
 mmatrixar =
 nomatrix.EpsilonElasticityMatrix[newsmatrix, nmatrix,
   vmatrix, p, Normalized -> False, opts].linkmatrix;
(*Eqn 5.25 b.*)
 coefmatrix =
 NewToOld[-linkmatrix.Inverse[mmatrixar].nomatrix,
   linkmatrixtransform];
```

```
(*Evaluate S with parameters and concentrations
   defined in steadystaterule and parameter rule.*)
 smatrixar = smatrix /. steadystaterule /. parameterrule;
```

```
(*Evaluate v with parameters and concentrations
   defined in steadystaterule and parameter rule.*)
vmatrixar = vmatrix /. steadystaterule /. parameterrule;
```

```
(*Assign the value of
   the Normalized Option to normalized.*)
normalized = Normalized /. {opts} /.
   Options[FluxControlMatrix];
```

```
(* Evaluate the control
   matrix with various normalization options.*)
If[normalized,
  (*If normalized=
    True calculate normalized matrix with Eqn 5.34 a.*)
```

```
      Inverse[DiagonalMatrix[smatrixar]].coefmatrix.
       DiagonalMatrix[vmatrixar] ,
      (*If normalized=False return non-normalized matrix*)
      coefmatrix,
      (*Otherwise calculate normalized matrix.*)
      Inverse[DiagonalMatrix[smatrixar]].
       coefmatrix.DiagonalMatrix[vmatrixar]
    ]
  ] /; (Length[smatrix] == Length[nmatrix] &&
      Length[vmatrix] == Length[Transpose[nmatrix]])  ||
Message[MetabolicControlAnalysis::badarg,
     "reactions and/or metabolites",
     "columns and/or rows"](*Error Message.*)


(*ConcResponseMatrix.*)
ConcResponseMatrix[smatrix_List, nmatrix_List,
  vmatrix_List, partp_List, p_List: {}, opts___Rule] :=

 Module[{coefmatrix, smatrixar, partpar, normalized,
    parameterrule, pars, parvals, steadystaterule},

   (*Calculate the un-
     normalized response matrix using Eqn 5.28.*)
   coefmatrix = ConcControlMatrix[smatrix, nmatrix, vmatrix, p,
      Normalized → False, opts].PiElasticityMatrix[smatrix,
      nmatrix, vmatrix, partp, p, Normalized → False, opts];

   (*Convert parameter list into a
     replacement rule called parameter rule.*)
   If[p == {},
   parameterrule = {},
   pars = Transpose[p][[1]]; parvals = Transpose[p][[2]];
   parameterrule =
     Table[pars[[i]] -> parvals[[i]], {i, Length[pars]}]
   ];

   (*Create the steady state replacement rule.*)
   steadystaterule =
    SteadyStateConc /. {opts} /. Options[ConcControlMatrix];

   (*Evaluate S with parameters and concentrations
     defined in steadystaterule and parameter rule.*)
     smatrixar = smatrix /. steadystaterule /. parameterrule;
```

```
(*Evaluate partp with parameter rule.*)
partpar = partp /. steadystaterule /. parameterrule;

(*Assign the value of
  the Normalized Option to normalized.*)
 normalized = Normalized /. {opts} /.
  Options[ConcControlMatrix];

(* Evaluate the response
  matrix with various normalization options.*)
If[normalized,
  (*If normalized=True calculate normalized matrix.*)
  Inverse[DiagonalMatrix[smatrixar]].
   coefmatrix.DiagonalMatrix[partpar] ,
  (*If normalized=False return non-normalized matrix.*)
  coefmatrix,
  (*Otherwise calculate normalized matrix.*)
  Inverse[DiagonalMatrix[smatrixar]].
   coefmatrix.DiagonalMatrix[partpar]
 ]
 ]


(*Conservation Relations.*)
(*This algorithm detects non-
  negative conservation relations using convex
   analysis. The algorithm is based on a method described
   in Heinrich and Schuster (1996) Section 3.1.2.*)
ConservationRelations[smatrix_, nmatrix_, opts___Rule] :=

 Module[{lhstableau, rhstableau, tableau, mc,
   mr, g, columnindicestable, xx, nn, intersection,
   conditions, gmatrix, gmatrixrule, i, tn, k, l},
  (*Create initial tableau; (N I).*)
  lhstableau[0] = nmatrix;
  rhstableau[0] = IdentityMatrix[Length[nmatrix]];
   tableau[0] = AppendRows[nmatrix, rhstableau[0]];

  (*Number of rows and columns in lhstableau.*)
  mc = Dimensions[lhstableau[0]][[2]];
  mr = Length[lhstableau[0]];

  (*Next Do statement contains a subroutine
     for determining a series of tableau. First
     time through the Do statement we start with
```

tableau[0] and finish with tableau[1]. This
is done by taking all possible (allowable) non-
negative linear combinations of pairs of rows
in tableau[0] that give a zero in the first
column. Next time through we do the same starting
with tableau[1] so that tableau[2] has only
zeros in columns 1 *and* 2. The Do statement ends
when tableau[mc=number of columns in lhs tableau]
has been created;
this will have a lhs tableau of zeros.*)

Do[
  (*Create a table of column indices for tableau[tn]. Each
    row of this table contains the positions of
    zero elements in the corresponding tableau row;
   note only 0 s in rhstableau are considered;
   see Eqn 3.25 for a definition of these indices. This
    table will be used later in the program.*)
  columnindicestable[tn] = Table[
    Flatten[Position[rhstableau[tn][[i]], 0]] + mc,
    {i, Length[tableau[tn]]}];

  g = 0; (*Set the counter, g, to 0 each time through the Do
    statement. g will count the number of allowable
    positive linear combinations in each tableau.*)

  (*The following For statement is created to count through
    the rows in the current tableau, tableau[i].*)
  For[i = 1, i < Length[tableau[tn]] + 1, i++,

    (*Check to see if element tn+1 in row i not 0.*)
    (*If[1].*)
    If[tableau[tn][[i, tn + 1]] != 0,
      (*If this element is not zero then use the following
        For statement to count through the rows below
        row i. This For statement will be used to
        create all possible (allowable) positive linear
        combinations of row i with those rows below row i.*)
      For[k = i + 1, k < Length[tableau[tn]] + 1, k++,

        (*An allowable positive linear combination of row i
          with row k will have to satisfy 2 conditions;
          Eqn 3.27 (Condition 1) and Eqn 3.28 (Condition 2).*)

```
(*START CHECK CONDITIONS.*)
(*If[2].*)
(*CHECK CONDITION1.*)
If[
 tableau[tn][[i, tn + 1]] * tableau[tn][[k, tn + 1]] < 0 ,
 (*If condition 1 is satisfied we need
   to then check condition 2*)
 (*CHECK CONDITION 2.*)
 (*First determine a table
   of columnindices which does not contain
   the column indices for rows i and k.*)
 xx[tn] = Delete[columnindicestable[tn], {{i}, {k}}];
 (*assign intersection = intersection of column
     indices of rows i and k. ie LHS of Eqn 3.28.*)
 intersection = Intersection[columnindicestable[
     tn][[i]], columnindicestable[tn][[k]]];
 (*If[3].*)
 If[xx[tn] == {},
  (*If xx[tn] ={} condition 2 is satisfied.*)
  conditions = True,
  (*Otherwise*)
  For[l = 1, l < Length[tableau[tn]] - 1, l++,
   (*Check Eqn 3.28 (Condition 2) for row l.*)
   (*If[4].*)
   If[Complement[intersection, xx[tn][[l]]] == {},
    (*If it is not satisfied set conditions=
      False and break from the current For loop.*)
    conditions = False; Break[],
    (*Otherwise set conditions=True and
       continue checking other rows in xx[tn].*)
    conditions = True
   ];(*End If[4].*)
  ];(*End of l++*)
 ];(*End If[3].*)
];(*End If[2].*)
(*END CHECK CONDITIONS.*)

(*If conditions are satisfied, conditions =
   True and we can use Eqn 3.26 to calculate a
   valid non negative linear combination.*)
(*If[5].*)
If[conditions,
 g++;
 nn[tn, g] =
```

```
        Abs[tableau[tn][[i, tn + 1]]] * tableau[tn][[k]] +
          Abs[tableau[tn][[k, tn + 1]]] * tableau[tn][[i]];
        conditions = False
      ]; (*End If[5].*)
     ], (*End of k++*)
     (*False part of If[1]; If the tested element is 0,
      row i goes into the new tableau.*)
     g++;
     nn[tn, g] = tableau[tn][[i]]
    ];(*End of If[1].*)
   ]; (*End of i++*)

   (*Create new tableau.*)
   tableau[tn + 1] = Table[nn[tn, i], {i, g}];
   rhstableau[tn + 1] = TakeColumns[tableau[tn + 1], -mr]
   , {tn, 0, mc - 1}](*End of Do.*);

(*Construct G matrix from the
  rhs of the last tableau (tableau[mc]).*)
If[tableau[mc] == {},
 gmatrix = {},
 gmatrix = Table[Flatten[RowReduce[{rhstableau[mc][[i]]}]],
   {i, Length[rhstableau[mc]]}]
];

(*Assign True or False to gmatrixrule
  depending of the value of the GMatrix Option.*)
gmatrixrule = GMatrix /. {opts} /.
  Options[ConservationRelations2];

(*Returns G or conservation relations,
 depending on value of gmatrixrule.*)
If[gmatrix == {},
 (*If gmatrix ={}, print {}*)
 {},
 (*Otherwise*)
 If[gmatrixrule,
  (*If gmatrixrule = True return gmatrix.*)
  gmatrix,
  (*If gmatrixrule =
    False return conservation relations.*)
  Print[gmatrix.smatrix, " == ",
   Array[Global`Const, Length[gmatrix]]]
 ]
```

```
    ]
  ];

(*EpsilonElasticityMatrix.*)
EpsilonElasticityMatrix[smatrix_List,
  nmatrix_List, vmatrix_List, p_List: {}, opts___Rule] :=

 Module[{parameterrule, pars, parvals, normalized,
    epsilonelasticitymatrix, vmatrixar, smatrixar},

  (*Convert parameter list into a
    replacement rule called parameter rule.*)
  If[p == {},
  parameterrule = {},
  pars = Transpose[p][[1]]; parvals = Transpose[p][[2]];
  parameterrule =
    Table[pars[[i]] -> parvals[[i]], {i, Length[pars]}]
  ];

  (*Create the steady state replacement rule.*)
   steadystaterule =
   SteadyStateConc /. {opts} /. Options[ConcControlMatrix];

  (*Calculate the un-
    normalized epsilon elasticity matrix using Eqn 5.35 a.*)
  epsilonelasticitymatrix = Outer[D, vmatrix, smatrix] /.
    steadystaterule /. parameterrule;

  (*Evaluate v with parameters and concentrations
    defined in steadystaterule and parameter rule.*)
  vmatrixar = vmatrix /. steadystaterule /. parameterrule;

  (*Evaluate S with parameters and concentrations
    defined in steadystaterule and parameter rule.*)
   smatrixar = smatrix /. steadystaterule /. parameterrule;

  (*Assign the value of
    the Normalized Option to normalized.*)
  normalized = Normalized /. {opts} /.
    Options[EpsilonElasticityMatrix];

  (*Evaluate the elasticity
    matrix with various normalization options.*)
  If[normalized,
```

```
(*If normalized=
  True calculate normalized matrix with Eqn 5.33 a*)
Inverse[DiagonalMatrix[vmatrixar]].
  epsilonelasticitymatrix.DiagonalMatrix[smatrixar],
(*If normalized=False return non-normalized matrix*)
epsilonelasticitymatrix,
(*Otherwise delete the matrix rows defined in normalized
  before normalizing the matrix. i.e. this option allows
  the removal of zero fluzes before normalization.*)
Inverse[DiagonalMatrix[Delete[vmatrixar, normalized]]].
  Delete[epsilonelasticitymatrix, normalized].
  DiagonalMatrix[smatrixar]
 ]
]


(*FluxControlMatrix.*)
FluxControlMatrix[smatrix_List, nmatrix_List,
  vmatrix_List, p_List: {}, opts___Rule] :=

Module[{linkmatrixsolution, newsmatrix,
    linkmatrix, nomatrix, linkmatrixtransform,
    paramterrule, pars, parvals, steadystaterule,
    mmatrixar, coefmatrix, vmatrixar, normalized},

  (*Rearrange S, N, v into Eqn 3.8
    using LinkMatrix.*)
    linkmatrixsolution = LinkMatrix[smatrix, nmatrix];
  newsmatrix = linkmatrixsolution[[1, 2]];
  linkmatrix = linkmatrixsolution[[2, 2]];
  nomatrix = linkmatrixsolution[[3, 2]];
  linkmatrixtransform = linkmatrixsolution[[6, 2]];

  (*Convert parameter list into a
    replacement rule called parameter rule.*)
  If[p == {},
  parameterrule = {},
  pars = Transpose[p][[1]]; parvals = Transpose[p][[2]];
  parameterrule =
    Table[pars[[i]] -> parvals[[i]], {i, Length[pars]}]
  ];

  (*Create the steady state replacement rule.*)
  steadystaterule =
  SteadyStateConc /. {opts} /. Options[ConcControlMatrix];
```

```
(*Calculate the non-normalized flux
   control matrix using Eqn 5.26 b and Eqn 5.24.*)
(*Eqn 5.24.*)
mmatrixar =
 nomatrix.EpsilonElasticityMatrix[newsmatrix, nmatrix,
   vmatrix, p, Normalized -> False, opts].linkmatrix;
(*Eqn 5.26 b.*)
coefmatrix = IdentityMatrix[Length[Transpose[nomatrix]]] -
   EpsilonElasticityMatrix[newsmatrix, nmatrix,
     vmatrix, p, Normalized -> False, opts].
   linkmatrix.Inverse[mmatrixar].nomatrix;

(*Evaluate v with parameters and concentrations
   defined in steadystaterule and parameter rule.*)
 vmatrixar = vmatrix /. steadystaterule /. parameterrule;

(*Assign the value of
   the Normalized Option to normalized.*)
 normalized = Normalized /. {opts} /.
   Options[FluxControlMatrix];

(* Evaluate the control
   matrix with various normalization options.*)
If[normalized,
 (*If normalized=
   True calculate normalized matrix with Eqn 5.34 b*)
 Inverse[DiagonalMatrix[vmatrixar]].coefmatrix.
  DiagonalMatrix[vmatrixar],
 (*If normalized=False return non-normalized matrix*)
 coefmatrix,
 (*Otherwise delete the matrix rows defined in normalized
   before normalizing the matrix. i.e., this option allows
   the removal of zero fluxes before normalization.*)
 Inverse[DiagonalMatrix[Delete[vmatrixar, normalized]]].
  Delete[coefmatrix, normalized].DiagonalMatrix[vmatrixar]
 ]
] /; (Length[smatrix] == Length[nmatrix] &&
  Length[vmatrix] == Length[Transpose[nmatrix]]) ||
Message[MetabolicControlAnalysis::badarg,
  "reactions and/or metabolites",
  "columns and/or rows"](*Error Message.*)
```

```
(*FluxResponseMatrix.*)
FluxResponseMatrix[smatrix_List, nmatrix_List,
  vmatrix_List, partp_List, p_List: {}, opts___Rule] :=

Module[{coefmatrix, vmatrixar, partpar, normalized,
  parameterrule, pars, parvals, steadystaterule},

  (*Calculate the non-normalized matrix with Eqn 5.29*)
  coefmatrix = FluxControlMatrix[smatrix, nmatrix, vmatrix, p,
    Normalized → False, opts].PiElasticityMatrix[smatrix,
    nmatrix, vmatrix, partp, p, Normalized → False, opts];

  (*Convert parameter list into a
    replacement rule called parameter rule.*)
  If[p == {},
  parameterrule = {},
  pars = Transpose[p][[1]]; parvals = Transpose[p][[2]];
  parameterrule =
    Table[pars[[i]] -> parvals[[i]], {i, Length[pars]}]
  ];

  (*Create the steady state replacement rule.*)
  steadystaterule =
   SteadyStateConc /. {opts} /. Options[ConcControlMatrix];

  (*Evaluate v with parameters and concentrations
    defined in steadystaterule and parameter rule.*)
   vmatrixar = vmatrix /. steadystaterule /. parameterrule;

  (*Evaluate partp with parameter rule.*)
  partpar = partp /. steadystaterule /. parameterrule;

  (*Assign the value of
    the Normalized Option to normalized.*)
  normalized = Normalized /. {opts} /.
    Options[FluxControlMatrix];

  (*Evaluate the response
    matrix with various normalization options.*)
  If[normalized,
    (*If normalized=True calculate normalized matrix*)
    Inverse[DiagonalMatrix[vmatrixar]].
    coefmatrix.DiagonalMatrix[partpar],
    (*If normalized=False return non-normalized matrix*)
```

```
    coefmatrix,
    (*Otherwise delete the matrix rows defined in normalized
       before normalizing the matrix. i.e., this option allows
       the removal of zero fluxes before normalization.*)
    Inverse[DiagonalMatrix[Delete[vmatrixar, normalized]]].
      Delete[coefmatrix, normalized].DiagonalMatrix[partpar]
  ]
]


(*LinkMatrix.*)
(*Rearranges S, N, v into Eqn 3.8.*)
LinkMatrix[smatrix_List, nmatrix_List] :=

Module[{linkmatrixtransform, newnmatrix,
    newgmatrix, nomatrix, linkdashmatrix, linkmatrix,
    newsmatrix, modifiedgmatrix, rgm, nrgm},

linkmatrixtransform = LinkMatrixTransform[nmatrix];
newnmatrix = OldToNew[nmatrix, linkmatrixtransform];
newgmatrix = NullSpace[Transpose[newnmatrix]]
    (*See Eqn 3.5.*);
nomatrix = TakeRows[newnmatrix,
    Length[newnmatrix] - Length[newgmatrix]];

    (*Rearrange newgmatrix calculated above into
       the form given in Eqn 3.11. i.e., G=(-L' I).*)
    (*Determine L'.*)
    If[newgmatrix == {},
      linkdashmatrix = {},
     (*Otherwise.*)
     (*Cycle the elements of each row in newgmatrix to
        the left by the number of rows in nomatrix.*)
     rgm = Table[RotateLeft[newgmatrix[[i]], Length[nomatrix]],
        {i, Length[newgmatrix]}];
     (*Use RowReduce to rearrange to (I -L').*)
     nrgm = RowReduce[rgm];

     (*Delete I in (I -L') to give L'.*)
     deletetable = Flatten[Table[{i, j},
        {i, Length[newgmatrix]}, {j, Length[newgmatrix]}], 1];
     linkdashmatrix = -Delete[nrgm, deletetable]
    ];

    (*G=(-L',I).*)
```

```
modifiedgmatrix = Table[Flatten[{-linkdashmatrix[[i]],
    IdentityMatrix[Length[linkdashmatrix]][[i]]}],
  {i, Length[linkdashmatrix]}];


(*L.*)
linkmatrix = Flatten[
  {IdentityMatrix[Length[nomatrix]], linkdashmatrix }, 1];


(*new S.*)
newsmatrix =
 OldToNew[Transpose[{smatrix}], linkmatrixtransform];


(*Output Matrix.*)
{{"S̄", Flatten[newsmatrix] },
{"L̄", linkmatrix },
{"N̄⁰ ", nomatrix },
{"N̄", newnmatrix },
{"G", modifiedgmatrix },
 {"LinkMatrixTransform", linkmatrixtransform}}


 ] /; Length[smatrix] == Length[nmatrix]  ||
  Message[MetabolicControlAnalysis::badarg,
    "reactions and/or metabolites",
    "columns and/or rows"](*Error message.*)

(*MMatrix.*)
MMatrix[smatrix_List, nmatrix_List,
  vmatrix_List, p_List: {}, opts___Rule] :=


(*See Eqn 2.83 for the definition of the MMatrix.*)
nmatrix.
  EpsilonElasticityMatrix[smatrix, nmatrix, vmatrix, p, opts]


(*NDSolveMatrix.*)
NDSolveMatrix[smatrix_List, nmatrix_List,
  vmatrix_List, initialconditions_List,
  timerange_List, p_List: {}, opts___Rule] :=


Module[{equations, pars, parvals,
    nvmatrix, parameterrule, removet, smatrixnot},


(*Convert parameter list into a
    replacement rule called parameter rule.*)
  If[p == {},
```

```
parameterrule = {},
pars = Transpose[p][[1]]; parvals = Transpose[p][[2]];
parameterrule =
   Table[pars[[i]] -> parvals[[i]], {i, Length[pars]}]
];

(*Define the list of metabolites for which to apply
   NDSolve. i.e., remove [t] index from each metabolite.*)
removet[x_[Global`t]] := x;
smatrixnot = Map[removet, smatrix];

(*Evaluate v with parameter rule.*)
nvmatrix = vmatrix /. parameterrule;

(*Express the system of ordinary differential
   equations that are described by S, N, V,
   and p in a form that can be used by NDSolve*)
equations = Table[∂Global`t smatrix[[i]] ==
   (nmatrix .nvmatrix)[[i]], {i, Length[smatrix]}];

(*Apply NDSolve including the Options, opts.*)
NDSolve[Flatten[{equations , initialconditions}],
   smatrixnot, timerange, opts]

] /; (Length[smatrix] == Length[nmatrix] &&
   Length[vmatrix] == Length[Transpose[nmatrix]]) ||
Message[MetabolicControlAnalysis::badarg,
   "reactions and/or metabolites",
   "columns and/or rows"](*Error Message*)

(*NMatrix.*)
NMatrix[eqn_List, ext_List: {}] :=

Module[{modifiedeqn0, deletetable,
   modifiedeqn1, modifiedeqn2, smatrix, nmatrix},

(*Remove reaction labels and create
   modified equation matrix from the list eqn.*)
modifiedeqn0 = Table[{-eqn[[i, Length[eqn[[i]]], 1]] +
      eqn[[i, Length[eqn[[i]]], 2]]}, { i, Length[eqn]}];

(*Delete any external parameters from the matrix,
   modifiedeqn0.*)
deletetable = Flatten[Table[Position[modifiedeqn0, ext[[i]]],
```

```
        {i, Length[ext]}], 1];
modifiedeqn1 = Delete[modifiedeqn0, deletetable];

(*Generate substrate matrix.*)
 modifiedeqn2 = Flatten[modifiedeqn1 /. Plus -> List];
 smatrix =
     Union[Delete[modifiedeqn2,
       Position[modifiedeqn2, (x_ /; NumberQ[x]), 2 ]]];

(*Generate nmatrix.*)
nmatrix =
   Table[
      If[Position[modifiedeqn1[[j]], smatrix[[i]]] == {},
         0,
         Coefficient[modifiedeqn1[[j]], smatrix[[i]]][[1]]
         ],
      {i, Length[smatrix]}, {j, Length[modifiedeqn1]}]
]

(*NSteadyState.*)
NSteadyState[smatrix_List, nmatrix_List, vmatrix_List,
  p_List: {}, initialvalues_List, opts___Rule] :=

 Module[
    {i, linkmatrixsolution, newsmatrix, constmatrix,
     sbmatrix, samatrix, nomatrix, newgmatrix, rhs,
     lhs, dmrule, newvmatrix, ss, parameterrule, pars, parvals,
     crrule, newinitialvaluerule, indepinitialvaluerule,
     indepinitialvaluelist, indepinitialvalueseq},

    (*Rearrange S, N, v into Eqn 3.8
      using LinkMatrix.*)
    linkmatrixsolution = LinkMatrix[smatrix, nmatrix];
    newsmatrix = linkmatrixsolution[[1, 2]];
    nomatrix = linkmatrixsolution[[3, 2]];
    newgmatrix = linkmatrixsolution[[5, 2]]; (*Eqn 3.11.*)
    samatrix = Take[newsmatrix, Length[nomatrix]];
    sbmatrix = Take[newsmatrix, -Length[newgmatrix]];

    (*Generate a replacement rule using Eqn 3.10
      so that each concentration in sbmatrix can be
      expressed as a function of conservation constants
      and the independent metabolite concentrations.*)
    constmatrix = Array[Global`Konst, Length[newgmatrix]];
```

```
rhs = constmatrix -
  TakeColumns[newgmatrix, Length[nomatrix]].samatrix ;
lhs = TakeColumns[newgmatrix, -Length[newgmatrix]].sbmatrix;
dmrule =
  Table[lhs[[i]] -> rhs[[i]], {i, Length[newgmatrix]}] ;

(*Convert parameter list into
   a replacement rule called parameter rule.*)
  If[p == {},
parameterrule = {},
pars = Transpose[p][[1]]; parvals = Transpose[p][[2]];
parameterrule =
    Table[pars[[i]] -> parvals[[i]], {i, Length[pars]}]
  ];

(*Generate a replacement
   rule so that the constants, Konst[i],
  can be expressed in terms of the constants, Const[i],
  that were determined using ConservationRelations.*)
  crrule = ConservationRelationsTransform[smatrix, nmatrix];

(*Generate a new vmatrix using the parameter replacement
   rule, the dependent metabolite replacement rule,
  and the constants replacement rule. The new
   vmatrix has rate expressions in terms of the
   independent metabolite concentrations and
   non negative conservation constants only.*)
  newvmatrix = vmatrix /. dmrule /. parameterrule /. crrule;

(*Convert the steadystate estimate replacement rule list,
   initialvalues, into a form that can be
   used in FindRoot. And only include the
   independent metabolites in this sequence.*)
  newinitialvaluerule = OldToNew[Transpose[{initialvalues}],
    linkmatrixsolution[[6, 2]]];
  indepinitialvaluerule = Take[newinitialvaluerule,
    Length[nomatrix]];
  indepinitialvaluelist =
  Table[{indepinitialvaluerule[[i, 1, 1]],
      indepinitialvaluerule[[i, 1, 2]]},
    {i, Length[nomatrix]}];
  indepinitialvalueseq = ReplacePart[
    indepinitialvaluelist, Sequence, 0];
```

```
(*Use FindRoot to solve
   Eqn 2.9 for the independent metabolites.*)
ss = FindRoot[nomatrix .newvmatrix ==
     Table[0, {i, Length[samatrix]}],
   Evaluate[indepinitialvalueseq], opts];


(*Return the values of both the independent
   and dependent metabolite concentrations.*)
Union[ss, dmrule /. ss /. crrule]


] /; (Length[smatrix] == Length[nmatrix] &&
       Length[vmatrix] == Length[Transpose[nmatrix]]) ||
Message[MetabolicControlAnalysis::badarg,
    "reactions and/or metabolites",
       "columns and/or rows"](*Error message.*)


(*PartialConcResponse.*)
PartialConcResponse[smatrix_List, nmatrix_List, vmatrix_List,
  n_Integer, partp_List, p_List: {}, opts___Rule] :=

 Module[{coefmatrix, parameterrule, pars,
   parvals, steadystaterule, smatrixar,  partpar },

  (*Calculate the non-normalized
     partial response coefficients. See Eqn 5.30.*)
  coefmatrix = ConcControlMatrix[smatrix, nmatrix,
       vmatrix, p, Normalized → False, opts][[n]] *
    PiElasticityMatrix[smatrix, nmatrix, vmatrix,
       partp, p, Normalized → False, opts];

  (*Convert parameter list into a
    replacement rule called parameter rule.*)
  If[p == {},
  parameterrule = {},
  pars = Transpose[p][[1]]; parvals = Transpose[p][[2]];
  parameterrule =
    Table[pars[[i]] -> parvals[[i]], {i, Length[pars]}]
  ];

(*Create the steady state replacement rule.*)
steadystaterule =
    SteadyStateConc /. {opts} /. Options[ConcControlMatrix];

  (*Evaluate S with parameters and concentrations
```

```
      defined in steadystaterule and parameter rule.*)
   smatrixar = smatrix /. steadystaterule /. parameterrule;

(*Evaluate partp with parameter rule.*)
  partpar = partp /. parameterrule;

  (* Evaluate partial response
    matrix with various normalization options.*)
  normalized = Normalized /. {opts} /.
    Options[FluxControlMatrix];

If[normalized,
    (*If normalized=True calculate normalized matrix*)
    1 / smatrixar[[n]] * coefmatrix.DiagonalMatrix[partpar] ,
    (*If normalized=False return non-normalized matrix*)
    coefmatrix,
    (*Otherwise calculate the normalized matrix*)
    1 / smatrixar[[n]] * coefmatrix.DiagonalMatrix[partpar]
    ]
  ]


(*PartialFluxResponse.*)
PartialFluxResponse[smatrix_List, nmatrix_List, vmatrix_List,
  n_Integer, partp_List, p_List: {}, opts___Rule] :=

 Module[{coefmatrix, parameterrule, pars,
    parvals, steadystaterule, smatrixar, partpar },

  (*Calculate the non-
    normalized partial response coefficients. See Eqn 5.30.*)
coefmatrix = FluxControlMatrix[smatrix, nmatrix,
      vmatrix, p, Normalized → False, opts][[n]] *
    PiElasticityMatrix[smatrix, nmatrix, vmatrix,
      partp, p, Normalized → False, opts];

  (*Convert parameter list into a
    replacement rule called parameter rule.*)
  If[p == {},
  parameterrule = {},
  pars = Transpose[p][[1]]; parvals = Transpose[p][[2]];
  parameterrule =
    Table[pars[[i]] -> parvals[[i]], {i, Length[pars]}]
  ];
```

```
(*Create the steady state replacement rule.*)
steadystaterule =
   SteadyStateConc /. {opts} /. Options[ConcControlMatrix];


  (*Evaluate v with parameters and concentrations
    defined in steadystaterule and parameter rule.*)
vmatrixar = vmatrix /. steadystaterule /. parameterrule;


  (*Evaluate partp with parameter rule.*)
partpar = partp /. parameterrule;

  (* Evaluate partial response
    matrix with various normalization options.*)
  normalized = Normalized /. {opts} /.
    Options[FluxControlMatrix];


If[normalized,
    (*If normalized=True calculate normalized matrix.*)
   1 / vmatrixar[[n]] * coefmatrix.DiagonalMatrix[partpar] ,
    (*If normalized=False return non-normalized matrix.*)
    coefmatrix,
    (*Otherwise calculate the normalized matrix.*)
    1 / vmatrixar[[n]] * coefmatrix.DiagonalMatrix[partpar]
  ]
 ]

(*PartialInternalConcResponse.*)
PartialInternalConcResponse[
   smatrix_List, nmatrix_List, vmatrix_List,
   n_Integer, m_Integer, p_List: {}, opts___Rule] :=

ConcControlMatrix[smatrix, nmatrix,
    vmatrix, p, Normalized → False, opts][[n]] *
  EpsilonElasticityMatrix[smatrix, nmatrix, vmatrix,
    p, Normalized → False, opts][[All, m]]

(*PartialInternalFluxResponse.*)
PartialInternalFluxResponse[
   smatrix_List, nmatrix_List, vmatrix_List,
   n_Integer, m_Integer, p_List: {}, opts___Rule] :=

FluxControlMatrix[smatrix, nmatrix,
    vmatrix, p, Normalized → False, opts][[n]] *
  EpsilonElasticityMatrix[smatrix, nmatrix, vmatrix,
```

```
     p, Normalized → False, opts] [[All, m]]

(*PiElasticityMatrix.*)
PiElasticityMatrix[smatrix_List, nmatrix_List,
  vmatrix_List, partp_List, p_List: {}, opts___Rule] :=

Module[{parameterrule, pars, parvals, steadystaterule,
   normalized, pielasticitymatrix, vmatrixar, partpar},

(*Convert parameter list into a
    replacement rule called parameter rule.*)
If[p == {},
  parameterrule = {},
  pars = Transpose[p][[1]]; parvals = Transpose[p][[2]];
  parameterrule =
    Table[pars[[i]] -> parvals[[i]], {i, Length[pars]}]
];

(*Create the steady state replacement rule.*)
steadystaterule =
  SteadyStateConc /. {opts} /. Options[PiElasticityMatrix];

(*Calculate the non-normalized pi elasticity matrix
    (see Eqn 5.35 b) with parameters and concentrations
    defined in steadystaterule and parameter rule.*)
pielasticitymatrix = Outer[D, vmatrix, partp] /.
    steadystaterule /. parameterrule;

  (*Evaluate v with parameters and concentrations
    defined in steadystaterule and parameter rule.*)
vmatrixar = vmatrix /. steadystaterule /. parameterrule;

  (*Evaluate partp with parameter rule.*)
partpar = partp /. parameterrule;

(*Evaluate pi elasticity matrix
    with various normalization options.*)
normalized = Normalized /. {opts} /.
    Options[PiElasticityMatrix];

If[normalized,
   (*If normalized=
    True calculate normalized matrix with Eqn 5.33 b*)
  Inverse[DiagonalMatrix[vmatrixar]].
```

```
        pielasticitymatrix.DiagonalMatrix[partpar],
      (*If normalized=False return non-normalized matrix*)
    pielasticitymatrix,
      (*Otherwise delete the matrix rows defined in normalized
        before normalizing the matrix. i.e., this option allows
        the removal of zero fluxes before normalization.*)
      Inverse[DiagonalMatrix[Delete[vmatrixar, normalized]]].
      Delete[pielasticitymatrix, normalized].
      DiagonalMatrix[partpar]
  ]
  ]


(*SMatrix.*)
SMatrix[eqn_List, ext_List: {}] :=


Module[{modifiedeqn0, deletetable,
    modifiedeqn1, modifiedeqn2, smatrix},


(*Remove reaction labels
    and create modified equation matrix.*)
modifiedeqn0 = Table[{-eqn[[i, Length[eqn[[i]]], 1]] +
      eqn[[i, Length[eqn[[i]]], 2]]}, {i, Length[eqn]}];


(*Delete any external parameters from the matrix.*)
  deletetable = Flatten[Table[
      Position[modifiedeqn0, ext[[i]]], {i, Length[ext]}], 1];
  modifiedeqn1 = Delete[modifiedeqn0, deletetable];


(*Generate substrate matrix.*)
  modifiedeqn2 = Flatten[modifiedeqn1 /. Plus -> List];
  smatrix =
      Union[Delete[modifiedeqn2,
        Position[modifiedeqn2, (x_ /; NumberQ[x]), 2]]]
  ]


(*Stability.*)
Stability[smatrix_List, nmatrix_List,
  vmatrix_List, p_List: {}, opts___Rule] :=


Module[{eigenvalues},

  (*Calculate the eigenvalues of the non-
    normalized MMatrix (Jacobian).*)
  eigenvalues = Eigenvalues[MMatrix[smatrix, nmatrix,
```

```
             vmatrix, p, Normalized -> False, opts]];

     (*Test for stability.*)
     If[MemberQ[eigenvalues, z_ /; Re[z] > 0],
       (*If any eigenvalues
         are positive steady state is unstable.*)
         Print["Asymptotically Unstable"];
           Print[eigenvalues],
       (*If no eigenvalues are
         positive then test to see if any are 0.*)
         If[MemberQ[eigenvalues, z_ /; Re[z] == 0.],
             (*if some eigenvalues
           are 0 steady state cannot be determined.*)
             Print["Stability cannot be determined." ];
               Print[eigenvalues],
             (*Otherwise print steady state is stable.*)
               Print["Asymptotically Stable"];
                 Print[eigenvalues]
           ]
     ];


]


(*SteadyState.*)
SteadyState[smatrix_List,
   nmatrix_List, vmatrix_List, p_List: {}] :=

 Module[
     {linkmatrixsolution, newsmatrix, constmatrix,
       sbmatrix, samatrix, nomatrix, newgmatrix, rhs,
       lhs, dmrule, newvmatrix,
       ss, parameterrule, pars, parvals, crrule},

     (*Rearrange S, N, v into Eqn 3.8
       using LinkMatrix.*)
     linkmatrixsolution = LinkMatrix[smatrix, nmatrix];
     newsmatrix = linkmatrixsolution[[1, 2]];
     nomatrix = linkmatrixsolution[[3, 2]];
     newgmatrix = linkmatrixsolution[[5, 2]]; (*Eqn 3.11.*)
     samatrix = Take[newsmatrix, Length[nomatrix]];
     sbmatrix = Take[newsmatrix, -Length[newgmatrix]];

     (*Generate a replacement rule using Eqn 3.10
       so that each concentration in sbmatrix can be
```

```
      expressed as a function of conservation constants
      and the independent metabolite concentrations.*)
    constmatrix = Array[Global`Konst, Length[newgmatrix]] ;
    rhs = constmatrix -
      TakeColumns[newgmatrix, Length[nomatrix]].samatrix ;
    lhs = TakeColumns[newgmatrix, -Length[newgmatrix]].sbmatrix;
    dmrule =
      Table[lhs[[i]] -> rhs[[i]], {i, Length[newgmatrix]}];

    (*Convert parameter list into
      a replacement rule called parameter rule.*)
    If[p == {},
parameterrule = {},
pars = Transpose[p][[1]]; parvals = Transpose[p][[2]];
parameterrule =
      Table[pars[[i]] -> parvals[[i]], {i, Length[pars]}]
      ];

    (*Generate a replacement
      rule so that the constants, Konst[i],
      can be expressed in terms of the constants, Const[i],
      that were determined using ConservationRelations.*)
    crrule = ConservationRelationsTransform[smatrix, nmatrix];

    (*Generate a new vmatrix using the parameter replacement
      rule, the dependent metabolite replacement rule,
      and the constants replacement rule. The new
      vmatrix has rate expressions in terms of the
      independent metabolite concentrations and non-
      negative conservation constants only.*)
    newvmatrix = vmatrix /. dmrule /. parameterrule /. crrule;

    (*Solve Eqn 2.9 for the independent metabolites.*)
    ss = Solve[nomatrix.newvmatrix ==
      Table[0, {i, Length[samatrix]}], samatrix];

    (*Return the values of both the independent
      and dependent metabolite concentrations.*)
    Table[Union[ss[[i]], dmrule /. ss[[i]] /. crrule],
      {i, Length[ss]}]

    ] /; (Length[smatrix] == Length[nmatrix] &&
        Length[vmatrix] == Length[Transpose[nmatrix]]) ||
Message[MetabolicControlAnalysis::badarg,
```

```
            "reactions and/or metabolites",
               "columns and/or rows"](*Error message.*)


(*StoichiometryMatrix.*)
StoichiometryMatrix[eqn_List, ext_List: {}] :=


Module[{modifiedeqn0, deletetable,
      modifiedeqn1, modifiedeqn2, reactionlist,
      smatrix, nmatrix, prestoichmatrix},


(*Remove reaction labels and create
      modified equation matrix from the list eqn.*)
modifiedeqn0 = Table[{-eqn[[i, Length[eqn[[i]]], 1]] +
         eqn[[i, Length[eqn[[i]]], 2]]}, { i, Length[eqn]}];


(*Delete any external parameters from the matrix,
   modifiedeqn0.*)
deletetable = Flatten[Table[Position[modifiedeqn0, ext[[i]]],
      {i, Length[ext]}], 1];
modifiedeqn1 = Delete[modifiedeqn0, deletetable];


(*Generate reaction list.*)
If[Length[eqn[[1]]] == 1,
   reactionlist = Prepend[Table[i , {i, Length[eqn]}], "s⌇r"],
   reactionlist =
      Prepend[Table[eqn[[i, 1]], {i, Length[eqn]}], "s⌇r"]
];


(*Generate substrate matrix.*)
 modifiedeqn2 = Flatten[modifiedeqn1 /. Plus -> List];
 smatrix =
      Union[Delete[modifiedeqn2,
         Position[modifiedeqn2, (x_ /; NumberQ[x]), 2 ]]];

(*Generate nmatrix.*)
nmatrix =
   Table[
      If[Position[modifiedeqn1[[j]], smatrix[[i]]] == {},
          0,
          Coefficient[modifiedeqn1[[j]], smatrix[[i]]][[1]]
          ],
      {i, Length[smatrix]}, {j, Length[modifiedeqn1]}];


(*Generate stoichiometry matrix.*)
```

```
(*Add smatrix.*)
prestoichmatrix =
    Transpose[Prepend[Transpose[nmatrix], smatrix]];
(*Add reaction list.*)
Prepend[prestoichmatrix, reactionlist]
]

(*VMatrix.*)
VMatrix[eqn_List, ext_List: {}] :=

Module[{reactionlist},

If[Length[eqn[[1]]] == 1,
  reactionlist = Table[Global`v[i], {i, Length[eqn]}],
  reactionlist = Table[Global`v[eqn[[i, 1]]], {i, Length[eqn]}]
];

  reactionlist
]

 (*End the private context.*)
End[];


 (*Protect exported symbols.*)
Protect[ConcControlMatrix, ConcResponseMatrix,
  EpsilonElasticityMatrix, FluxControlMatrix,
  FluxResponseMatrix, LinkMatrix, MMatrix, NDSolveMatrix,
  NMatrix, ConservationRelations, NSteadyState,
  PartialConcResponse, PartialFluxResponse,
  PartialInternalConcResponse, PartialInternalFluxResponse,
  PiElasticityMatrix, SMatrix, Stability,
  SteadyState, StoichiometryMatrix, VMatrix];

 (*End the package context.*)
EndPackage[]

DumpSave["MetabolicControlAnalysis`"]
```

Appendix 3 - Rate Equations for Enzymes of the Human Erythrocyte

The following is a list of the rate equations which have been used to model the individual metabolic reactions of the red cell. Justification for the rate equations can be found in Mulquiney and Kuchel (1999), McIntyre et al. (1989), and Thorburn and Kuchel (1985). Definitions of reaction names and metabolite abbreviations are found in Fig. 7.1 Note all units are in the appropriate combinations of mol, L, and s.

A3.1 Glycolytic Enzymes

A3.1.1 Hexokinase

Parameters

```
Ki[hk, mgatp] = 1.0 * 10^-3;
Km[hk, mgatp] = 1.0 * 10^-3;
Ki[hk, glc] = 4.7 * 10^-5;
Ki[hk, glc6p] = 4.7 * 10^-5;
Ki[hk, mgadp] = 1.0 * 10^-3;
Km[hk, mgadp] = 1.0 * 10^-3;
Kdi[hk, bpg] = 4.0 * 10^-3;
Kdi[hk, glc16p2] = 30 * 10^-6;
Kdi[hk, glc6p] = 10.0 * 10^-6;
Kdi[hk, gsh] = 3.0 * 10^-3;
HK = 25 * 10^-9;
kcatf[hk] :=
```
$$\frac{180 * 1.662}{1 + (10^\wedge\text{-pH1[t]} / 10^\wedge\text{-7.02}) + (10^\wedge\text{-9.55} / 10^\wedge\text{-pH1[t]})};$$
```
kcatr[hk] :=
```
$$\frac{1.16 * 1.662}{1 + (10^\wedge\text{-pH1[t]} / 10^\wedge\text{-7.02}) + (10^\wedge\text{-9.55} / 10^\wedge\text{-pH1[t]})};$$

Rate Equation

$$hkrd := \left(1 + \frac{MgATP[t]}{Ki[hk, mgatp]} + \frac{Glc[t]}{Ki[hk, glc]} + \right.$$

$$\frac{MgATP[t]\,Glc[t]}{Ki[hk, glc]\,Km[hk, mgatp]} + \frac{MgADP[t]}{Ki[hk, mgadp]} +$$

$$\frac{Glc6P[t]}{Ki[hk, glc6p]} + \frac{MgADP[t]\,Glc6P[t]}{Ki[hk, glc6p]\,Km[hk, mgadp]} +$$

$$\frac{B23PG[t] * Glc[t]}{Kdi[hk, bpg]\,Ki[hk, glc]} + \frac{Glc16P2[t]\,Glc[t]}{Kdi[hk, glc16p2]\,Ki[hk, glc]} +$$

$$\left. \frac{Glc6P[t] * Glc[t]}{Kdi[hk, glc6p]\,Ki[hk, glc]} + \frac{GSH[t]\,Glc[t]}{Kdi[hk, gsh]\,Ki[hk, glc]} \right);$$

$$v[hk] := Vol_i * \frac{HK}{hkrd}$$

$$\left(\frac{kcatf[hk]\,Glc[t]\,MgATP[t]}{Ki[hk, glc] * Km[hk, mgatp]} - \frac{kcatr[hk]\,Glc6P[t]\,MgADP[t]}{Ki[hk, glc6p] * Km[hk, mgadp]} \right);$$

A3.1.2 Glucose Phosphate Isomerase

Parameters

```
Km[gpi, glc6p] = 1.81 * 10 ^ - 4;
Km[gpi, fru6p] = 7.1 * 10 ^ - 5;
GPI = 2.18 * 10 ^ - 7;
kcatf[gpi] = 1.47 * 10 ^ 3;
kcatr[gpi] = 1.76 * 10 ^ 3;
```

Rate Equation

$$gpird := \left(1 + \frac{Glc6P[t]}{Km[gpi, glc6p]} + \frac{Fru6P[t]}{Km[gpi, fru6p]} \right);$$

$$v[gpi] := \frac{GPI}{gpird} \left(\frac{kcatf[gpi]\,Glc6P[t]}{Km[gpi, glc6p]} - \frac{kcatr[gpi]\,Fru6P[t]}{Km[gpi, fru6p]} \right);$$

A3.1.3 Phosphofructokinase

Parameters

```
Km[pfk, fru6p] = 7.5 * 10 ^ - 5;
Km[pfk, mgatp] = 6.8 * 10 ^ - 5;
Km[pfk, fru16p2] = 5.0 * 10 ^ - 4;
Km[pfk, mgadp] = 5.4 * 10 ^ - 4;
KT[pfk, atp] = 100 * 10 ^ - 6;
KT[pfk, mg] = 4.0 * 10 ^ - 3;
KT[pfk, b23pg] = 5 * 10 ^ - 3;
```

```
KR[pfk, amp] = 300 * 10^-6;
KR[pfk, phos] = 30 * 10^-3;
KR[pfk, glc16p2] = 10.0 * 10^-3;
PFK = 1.1 * 10^-7;
kcatf[pfk] = 822;
kcatr[pfk] = 36;
Ka[pfk] = 10^-7.05;
n[pfk] = 5;
```

Rate Equation

$$L[pfk] := \left(\frac{10^{\wedge}-pH1[t]}{Ka[pfk]}\right)^{n[pfk]} \left(1 + \frac{ATP[t]}{KT[pfk, atp]}\right)^4$$

$$\left(1 + \frac{Mg[t]}{KT[pfk, mg]}\right)^4 \left(1 + \frac{B23PG[t]}{KT[pfk, b23pg]}\right)^4 \Bigg/$$

$$\left(\left(1 + \frac{Fru6P[t]}{Km[pfk, fru6p]} + \frac{Fru16P2[t]}{Km[pfk, fru16p2]}\right)^4\right.$$

$$\left(1 + \frac{AMP[t]}{KR[pfk, amp]}\right)^4 \left(1 + \frac{Phos[t]}{KR[pfk, phos]}\right)^4$$

$$\left.\left(1 + \frac{Glc16P2[t]}{KR[pfk, glc16p2]}\right)^4\right);$$

$$pfkrd := \left(1 + \frac{Fru6P[t]}{Km[pfk, fru6p]} + \right.$$

$$\frac{MgATP[t]}{Km[pfk, mgatp]} + \frac{Fru6P[t] \, MgATP[t]}{Km[pfk, fru6p] \, Km[pfk, mgatp]}$$

$$+ \frac{Fru16P2[t]}{Km[pfk, fru16p2]} + \frac{MgADP[t]}{Km[pfk, mgadp]} + $$

$$\left.\frac{Fru16P2[t] \, MgADP[t]}{Km[pfk, fru16p2] * Km[pfk, mgadp]}\right);$$

$$v[pfk] := \frac{PFK}{(1 + L[pfk]) \, pfkrd} \left(\frac{kcatf[pfk] \, Fru6P[t] \, MgATP[t]}{Km[pfk, fru6p] \, Km[pfk, mgatp]} - \right.$$

$$\left.\frac{kcatf[pfk] \, Fru16P2[t] * MgADP[t]}{Km[pfk, fru16p2] \, Km[pfk, mgadp]}\right);$$

A3.1.4 Aldolase

Parameters

```
Km[ald, fru16p2] = 7.1 * 10^-6;
Ki[ald, fru16p2] = 19.8 * 10^-6;
Km[ald, grnp] = 35 * 10^-6;
Ki[ald, grnp] = 11 * 10^-6;
```

```
Km[ald, grap] = 189 * 10 ^ -6;
Ki[ald, b23pg] = 1.5 * 10 ^ -3;
ALD = 0.37 * 10 ^ -6;
kcatf[ald] = 68;
kcatr[ald] = 234;
```

Rate Equation

$$
\text{aldrd} := \left(1 + \frac{\text{B23PG[t]} + \text{Mg\$B23PG[t]}}{\text{Ki[ald, b23pg]}} + \frac{\text{Fru16P2[t]}}{\text{Km[ald, fru16p2]}} + \right.
$$

$$
\frac{\text{Km[ald, grnp]} * \text{GraP[t]}}{\text{Km[ald, grap]} \, \text{Ki[ald, grnp]}}
$$

$$
\left(1 + \frac{\text{B23PG[t]} + \text{Mg\$B23PG[t]}}{\text{Ki[ald, b23pg]}} \right) + \frac{\text{GrnP[t]}}{\text{Ki[ald, grnp]}}
$$

$$
+ \frac{\text{Km[ald, grnp]} \, \text{Fru16P2[t]} \, \text{GraP[t]}}{\text{Ki[ald, fru16p2]} \, \text{Km[ald, grap]} \, \text{Ki[ald, grnp]}} +
$$

$$
\left. \frac{\text{GraP[t]} * \text{GrnP[t]}}{\text{Km[ald, grap]} \, \text{Ki[ald, grnp]}} \right);
$$

$$
\text{v[ald]} :=
$$
$$
\frac{\text{ALD}}{\text{aldrd}} \left(\frac{\text{kcatf[ald]} \, \text{Fru16P2[t]}}{\text{Km[ald, fru16p2]}} - \frac{\text{kcatr[ald]} \, \text{GrnP[t]} \, \text{GraP[t]}}{\text{Ki[ald, grnp]} \, \text{Km[ald, grap]}} \right);
$$

A3.1.5 Triosphosphate Isomerase

Parameters

```
Km[tpi, grap] = 446 * 10 ^ -6;
Km[tpi, grnp] = 162.4 * 10 ^ -6;
TPI = 1.14 * 10 ^ -6;
kcatf[tpi] = 14560;
kcatr[tpi] = 1280;
```

Rate Equation

$$
\text{tpird} := \left(1 + \frac{\text{GraP[t]}}{\text{Km[tpi, grap]}} + \frac{\text{GrnP[t]}}{\text{Km[tpi, grnp]}} \right);
$$

$$
\text{v[tpi]} := \frac{\text{TPI}}{\text{tpird}} \left(\frac{\text{kcatf[tpi]} \, \text{GraP[t]}}{\text{Km[tpi, grap]}} - \frac{\text{kcatr[tpi]} \, \text{GrnP[t]}}{\text{Km[tpi, grnp]}} \right);
$$

A3.1.6 Glyceraldehyde Phosphate Dehydrogenase

Parameters

```
Km[gapdh, nad] = 45 * 10 ^ -6;
Ki[gapdh, nad] = 45 * 10 ^ -6;
Km[gapdh, phos] = 3.16 * 10 ^ -3;
```

```
Ki[gapdh, phos] = 3.16 * 10^-3;
Km[gapdh, grap] = 95 * 10^-6;
Ki[gapdh, grap] :=
```
$$\frac{1.59 * 10^{-19} * 2.997}{1 + (10^{-pH1[t]} / 10^{-7.5}) + (10^{-10.0} / 10^{-pH1[t]})};$$
```
Kid[gapdh, grap] = 31 * 10^-6;
Km[gapdh, b13pg] = 0.671 * 10^-6;
Ki[gapdh, b13pg] :=
```
$$\frac{1.52 * 10^{-21} * 2.997}{1 + (10^{-pH1[t]} / 10^{-7.5}) + (10^{-10.0} / 10^{-pH1[t]})}$$
```
Kid[gapdh, b13pg] = 1 * 10^-6;
```
$$Km[gapdh, nadh] := 3.3 * 10^{-6} \left(\frac{10^{-7.2}}{10^{-pH1[t]}} \right);$$

$$Ki[gapdh, nadh] := 10 * 10^{-6} \left(\frac{10^{-7.2}}{10^{-pH1[t]}} \right);$$
```
GDH = 7.66 * 10^-6;

kcatf[gapdh] :=
```
$$\frac{232 * 2.997}{1 + (10^{-pH1[t]} / 10^{-7.5}) + (10^{-10.0} / 10^{-pH1[t]})};$$
```
kcatr[gapdh] :=
```
$$\frac{171 * 2.997}{1 + (10^{-pH1[t]} / 10^{-7.5}) + (10^{-10.0} / 10^{-pH1[t]})};$$

Rate Equation

```
gdhrd :=
```
$$\left(\frac{GraP[t]}{Ki[gapdh, grap]} \left(1 + \frac{GraP[t]}{Kid[gapdh, grap]} \right) + \frac{B13PG[t]}{Ki[gapdh, b13pg]} \right.$$

$$\left(1 + \frac{GraP[t]}{Kid[gapdh, grap]} \right) + \frac{Km[gapdh, b13pg]\, NADH[t]}{Ki[gapdh, b13pg]\, Km[gapdh, nadh]} +$$

$$\frac{Km[gapdh, grap]\, NAD[t]\, Phos[t]}{Ki[gapdh, grap]\, Km[gapdh, nad]\, Ki[gapdh, phos]} +$$

$$\frac{NAD[t]\, GraP[t]}{Ki[gapdh, nad]\, Ki[gapdh, grap]} +$$

$$\frac{Phos[t]\, GraP[t]}{Ki[gapdh, phos]\, Ki[gapdh, grap]} \left(1 + \frac{GraP[t]}{Kid[gapdh, grap]} \right) +$$

$$\frac{NAD[t]\, B13PG[t]}{Ki[gapdh, nad]\, Ki[gapdh, b13pg]} +$$

$$\frac{Km[gapdh, b13pg]\, Phos[t]\, NADH[t]}{Ki[gapdh, phos]\, Ki[gapdh, b13pg]\, Km[gapdh, nadh]} +$$

$$\frac{GraP[t]\, NADH[t]}{Ki[gapdh, grap]\, Ki[gapdh, nadh]} +$$

$$\frac{B13PG[t]\, NADH[t]}{Ki[gapdh, b13pg]\, Km[gapdh, nadh]} +$$

$$\frac{\text{NAD}[t]\ \text{Phos}[t]\ \text{GraP}[t]}{\text{Km}[\text{gapdh, nad}]\ \text{Ki}[\text{gapdh, phos}]\ \text{Ki}[\text{gapdh, grap}]}\ +$$

$$(\text{Km}[\text{gapdh, grap}]\ \text{NAD}[t]\ \text{Phos}[t]\ \text{B13PG}[t]) / (\text{Ki}[\text{gapdh, grap}]$$

$$\text{Km}[\text{gapdh, nad}]\ \text{Ki}[\text{gapdh, phos}]\ \text{Kid}[\text{gapdh, b13pg}]) +$$

$$\frac{\text{Phos}[t]\ \text{GraP}[t]\ \text{NADH}[t]}{\text{Ki}[\text{gapdh, phos}]\ \text{Ki}[\text{gapdh, grap}]\ \text{Ki}[\text{gapdh, nadh}]}\ +$$

$$(\text{Km}[\text{gapdh, b13pg}]\ \text{Phos}[t]\ \text{B13PG}[t]\ \text{NADH}[t]) /$$

$$(\text{Ki}[\text{gapdh, b13pg}]\ \text{Km}[\text{gapdh, nadh}]$$

$$\left. \text{Ki}[\text{gapdh, phos}]\ \text{Kid}[\text{gapdh, b13pg}]) \right);$$

$$\mathbf{v}[\text{gapdh}] :=$$
$$\frac{\text{GDH}}{\text{gdhrd}} \left(\frac{\text{kcatf}[\text{gapdh}]\ \text{GraP}[t]\ \text{NAD}[t]\ \text{Phos}[t]}{\text{Km}[\text{gapdh, nad}]\ \text{Ki}[\text{gapdh, phos}]\ \text{Ki}[\text{gapdh, grap}]} - \right.$$
$$\left. \frac{\text{kcatr}[\text{gapdh}]\ \text{B13PG}[t]\ \text{NADH}[t]}{\text{Ki}[\text{gapdh, b13pg}]\ \text{Km}[\text{gapdh, nadh}]} \right);$$

A3.1.7 Phosphoglycerate Kinase

Parameters

```
Km[pgk, mgadp] = 100 * 10^-6;
Ki[pgk, mgadp] = 80 * 10^-6;
Km[pgk, b13pg] = 2 * 10^-6;
Ki[pgk, b13pg] = 1.6 * 10^-6;
Km[pgk, mgatp] = 1 * 10^-3;
Ki[pgk, mgatp] = 0.186 * 10^-3;
Km[pgk, p3ga] = 1.1 * 10^-3;
Ki[pgk, p3ga] = 0.205 * 10^-3;
PGK = 2.74 * 10^-6;
kcatf[pgk] = 2290;
kcatr[pgk] = 917;
```

Rate Equation

$$\text{pgkrd} := \left(1 + \frac{\text{MgADP}[t]}{\text{Ki}[\text{pgk, mgadp}]} + \frac{\text{B13PG}[t]}{\text{Ki}[\text{pgk, b13pg}]} + \right.$$
$$\frac{\text{MgADP}[t]\ \text{B13PG}[t]}{\text{Ki}[\text{pgk, mgadp}]\ \text{Km}[\text{pgk, b13pg}]} + \frac{\text{MgATP}[t]}{\text{Ki}[\text{pgk, mgatp}]} +$$
$$\left. \frac{\text{P3GA}[t]}{\text{Ki}[\text{pgk, p3ga}]} + \frac{\text{MgATP}[t]\ \text{P3GA}[t]}{\text{Ki}[\text{pgk, mgatp}]\ \text{Km}[\text{pgk, p3ga}]} \right);$$

$$\mathbf{v}[\text{pgk}] := \frac{\text{PGK}}{\text{pgkrd}} \left(\frac{\text{kcatf}[\text{pgk}]\ \text{B13PG}[t]\ \text{MgADP}[t]}{\text{Ki}[\text{pgk, mgadp}]\ \text{Km}[\text{pgk, b13pg}]} - \right.$$
$$\left. \frac{\text{kcatr}[\text{pgk}]\ \text{P3GA}[t]\ \text{MgATP}[t]}{\text{Ki}[\text{pgk, mgatp}]\ \text{Km}[\text{pgk, p3ga}]} \right);$$

A3.1.8 Phosphoglycerate Mutase

Parameters

```
Km[pgm, p3ga] = 168 * 10 ^ -6;
Km[pgm, p2ga] = 25.6 * 10 ^ -6;
PGM = 410 * 10 ^ -9;
kcatf[pgm] = 0.795 * 10 ^ 3;
kcatr[pgm] = 0.714 * 10 ^ 3;
```

Rate Equation

$$\text{pgmrd} := \left(1 + \frac{\text{P3GA[t]}}{\text{Km[pgm, p3ga]}} + \frac{\text{P2GA[t]}}{\text{Km[pgm, p2ga]}} \right);$$

$$\text{v[pgm]} := \frac{\text{PGM}}{\text{pgmrd}} \left(\frac{\text{kcatf[pgm] P3GA[t]}}{\text{Km[pgm, p3ga]}} - \frac{\text{kcatr[pgm] P2GA[t]}}{\text{Km[pgm, p2ga]}} \right);$$

A3.1.9 Enolase

Parameters

```
Ki[eno, p2ga] = 140 * 10 ^ -6;
Km[eno, p2ga] = 140 * 10 ^ -6;
Ki[eno, pep] = 110.5 * 10 ^ -6;
Km[eno, pep] = 110.5 * 10 ^ -6;
Ki[eno, mg] = 46 * 10 ^ -6;
Km[eno, mg] = 46 * 10 ^ -6;
ENO = 0.22 * 10 ^ -6;
kcatf[eno] = 190;
kcatr[eno] = 50;
```

Rate Equation

$$\text{enord} := \left(1 + \frac{\text{P2GA[t]}}{\text{Ki[eno, p2ga]}} + \frac{\text{Mg[t]}}{\text{Ki[eno, mg]}} + \frac{\text{PEP[t]}}{\text{Ki[eno, pep]}} + \right.$$
$$\left. \frac{\text{P2GA[t]} * \text{Mg[t]}}{\text{Km[eno, p2ga] Ki[eno, mg]}} + \frac{\text{Mg[t]} * \text{PEP[t]}}{\text{Ki[eno, pep] Km[eno, mg]}} \right);$$

$$\text{v[eno]} :=$$
$$\frac{\text{ENO}}{\text{enord}} \left(\frac{\text{kcatf[eno] P2GA[t] Mg[t]}}{\text{Km[eno, p2ga] Ki[eno, mg]}} - \frac{\text{kcatr[eno] PEP[t] Mg[t]}}{\text{Ki[eno, pep] Km[eno, mg]}} \right);$$

A3.1.10 Pyruvate Kinase

Parameters

```
KT[pk, atp] = 3.39 * 10^-3;
KR[pk, pyr] = 2.0 * 10^-3;
KR[pk, pep] = 0.225 * 10^-3;
KR[pk, mgatp] = 3.0 * 10^-3;
KR[pk, mgadp] = 0.474 * 10^-3;
KR[f16p2] = 5.0 * 10^-6;
KR[g16p2] = 100 * 10^-6;
PK = 87.0 * 10^-9;
kcatf[pk] = 1386;
kcatr[pk] = 3.26;
```

Rate Equation

$$L[pk] := \left(\frac{10^{-6.8}}{10^{-pH1[t]}} \right) \left(1 + \frac{ATP[t]}{KT[pk, atp]} \right)^4 \Bigg/$$

$$\left(\left(1 + \frac{PEP[t]}{KR[pk, pep]} + \frac{Pyr[t]}{KR[pk, pyr]} \right)^4 \right.$$

$$\left. \left(1 + \frac{Fru16P2[t]}{KR[f16p2]} + \frac{Glc16P2[t]}{KR[g16p2]} \right)^4 \right);$$

$$pkrd :=$$

$$\left(1 + \frac{PEP[t]}{KR[pk, pep]} + \frac{MgADP[t]}{KR[pk, mgadp]} + \frac{PEP[t] * MgADP[t]}{KR[pk, pep] \, KR[pk, mgadp]} + \right.$$

$$\left. \frac{Pyr[t]}{KR[pk, pyr]} + \frac{MgATP[t]}{KR[pk, mgatp]} + \frac{Pyr[t] * MgATP[t]}{KR[pk, pyr] \, KR[pk, mgatp]} \right);$$

$$v[pk] := \frac{PK}{(1 + L[pk]) \, pkrd}$$

$$\left(\frac{kcatf[pk] \, PEP[t] * MgADP[t]}{KR[pk, pep] \, KR[pk, mgadp]} - \frac{kcatr[pk] \, Pyr[t] * MgATP[t]}{KR[pk, mgatp] \, KR[pk, pyr]} \right);$$

A3.1.11 Lactate Dehydrogenase

Parameters

```
Km[ldh, nadh] = 8.44 * 10^-6;
Ki[ldh, nadh] = 2.45 * 10^-6;
Km[ldh, nad] = 0.107 * 10^-3;
Ki[ldh, nad] = 0.503 * 10^-3;
```

$$Km[ldh, pyr] := 137 * 10^{-6} \left(\frac{1 + 10^{-6.8} / 10^{-pH1[t]}}{1 + 10^{-6.8} / 10^{-7.2}} \right);$$

$$\text{Ki[ldh, pyr]} := 228 * 10^-6 \left(\frac{1 + 10^-6.8 / 10^-\text{pH1[t]}}{1 + 10^-6.8 / 10^-7.2} \right);$$

$$\text{Kid[ldh, pyr]} = 0.101 * 10^-3;$$

$$\text{Km[ldh, lac]} := 1.07 * 10^-3 \left(\frac{1 + 10^-\text{pH1[t]} / 10^-6.8}{1 + 10^-7.2 / 10^-6.8} \right);$$

$$\text{Ki[ldh, lac]} := 7.33 * 10^-3 \left(\frac{1 + 10^-\text{pH1[t]} / 10^-6.8}{1 + 10^-7.2 / 10^-6.8} \right);$$

$$\text{LDH} = 3.43 * 10^-6;$$

$$\text{kcatf[ldh]} = 458;$$

$$\text{kcatr[ldh]} = 115;$$

Rate Equation

```
ldhrd :=
```

$$\left(\left(\left(1 + \frac{\text{Km[ldh, nadh] Pyr[t]}}{\text{Ki[ldh, nadh] Km[ldh, pyr]}} + \frac{\text{Km[ldh, nad] Lac[t]}}{\text{Km[ldh, lac] Ki[ldh, nad]}} \right) * \right.$$

$$\left(1 + \frac{\text{Pyr[t]}}{\text{Kid[ldh, pyr]}} \right) + \frac{\text{NADH[t]}}{\text{Ki[ldh, nadh]}} +$$

$$\frac{\text{NAD[t]}}{\text{Ki[ldh, nad]}} + \frac{\text{NADH[t] Pyr[t]}}{\text{Ki[ldh, nadh] Km[ldh, pyr]}} +$$

$$\frac{\text{Km[ldh, nad] NADH[t] Lac[t]}}{\text{Ki[ldh, nadh] Km[ldh, lac] Ki[ldh, nad]}} +$$

$$\frac{\text{Km[ldh, nadh] Pyr[t] NAD[t]}}{\text{Ki[ldh, nadh] Km[ldh, pyr] Ki[ldh, nad]}} +$$

$$\frac{\text{Lac[t] NAD[t]}}{\text{Km[ldh, lac] Ki[ldh, nad]}} +$$

$$\frac{\text{NADH[t] Pyr[t] Lac[t]}}{\text{Ki[ldh, nadh] Km[ldh, pyr] Ki[ldh, lac]}} +$$

$$\left. \frac{\text{Pyr[t] Lac[t] NAD[t]}}{\text{Ki[ldh, pyr] Km[ldh, lac] Ki[ldh, nad]}} \right);$$

$$\text{v[ldh]} := \frac{\text{LDH}}{\text{ldhrd}}$$

$$\left(\frac{\text{kcatf[ldh] Pyr[t] NADH[t]}}{\text{Ki[ldh, nadh] Km[ldh, pyr]}} - \frac{\text{kcatr[ldh] Lac[t] NAD[t]}}{\text{Ki[ldh, nad] Km[ldh, lac]}} \right);$$

A3.1.12 NADPH-Dependent Lactate Dehydrogenase

Parameters

$$\text{Km[ldhp, pyr]} = 4.14 * 10^-4;$$

$$\text{Km[ldhp, lac]} = 4.14 * 10^-4;$$

$$\text{kf[ldhp]} = 3.46 * 10^-3;$$

$$\text{kr[ldhp]} = 5.43 * 10^-7;$$

Rate Equation

$$\text{ldhprd} := 1 + \frac{\text{Pyr}[t]}{\text{Km}[\text{ldhp, pyr}]} + \frac{\text{Lac}[t]}{\text{Km}[\text{ldhp, lac}]} ;$$

$$v[\text{ldhp}] := \frac{1}{\text{ldhprd}}$$
$$\left(\frac{\text{kf}[\text{ldhp}] * \text{Pyr}[t] * \text{NADPH}[t]}{\text{Km}[\text{ldhp, pyr}]} - \frac{\text{kr}[\text{ldhp}] * \text{Lac}[t] * \text{NADP}[t]}{\text{Km}[\text{ldhp, lac}]} \right) ;$$

A3.2 Pentose Phosphate Pathway Enzymes

A3.2.1 Glucose 6-Phosphate Dehydrogenase

Parameters

```
k[g6pdh, 1] = 1.1 * 10^8;
k[g6pdh, 2] = 8.7 * 10^2;
k[g6pdh, 3] = 2.6 * 10^7;
k[g6pdh, 4] = 3.0 * 10^2;
k[g6pdh, 5] = 7.5 * 10^2;
k[g6pdh, 6] = 2.0 * 10^3;
k[g6pdh, 7] = 2.2 * 10^5;
k[g6pdh, 8] = 1.1 * 10^9;
k[g6pdh, 9] = 1.0 * 10^4;
k[g6pdh, 10] = 1.4 * 10^9;
G6PDH = 9.3 * 10^-8;
```

Rate Equation

```
N1[g6pdh] =
 k[g6pdh, 1] k[g6pdh, 3] k[g6pdh, 5] k[g6pdh, 7] k[g6pdh, 9];
N2[g6pdh] = k[g6pdh, 2] k[g6pdh, 4]
   k[g6pdh, 6] k[g6pdh, 8] k[g6pdh, 10];
D1[g6pdh] = k[g6pdh, 2] k[g6pdh, 9] (k[g6pdh, 4] k[g6pdh, 6] +
     k[g6pdh, 4] k[g6pdh, 7] + k[g6pdh, 5] k[g6pdh, 7]);
D2[g6pdh] =
  k[g6pdh, 1] k[g6pdh, 9] (k[g6pdh, 4] k[g6pdh, 6] +
     k[g6pdh, 4] k[g6pdh, 7] + k[g6pdh, 5] k[g6pdh, 7]);
D3[g6pdh] = k[g6pdh, 3] k[g6pdh, 5] k[g6pdh, 7] k[g6pdh, 9];
D4[g6pdh] = k[g6pdh, 2] k[g6pdh, 4] k[g6pdh, 6] k[g6pdh, 8];
D5[g6pdh] =
  k[g6pdh, 2] k[g6pdh, 10] (k[g6pdh, 4] k[g6pdh, 6] +
     k[g6pdh, 5] k[g6pdh, 6] + k[g6pdh, 5] k[g6pdh, 7]);
D6[g6pdh] = k[g6pdh, 1] k[g6pdh, 3]
   (k[g6pdh, 5] k[g6pdh, 7] + k[g6pdh, 5] k[g6pdh, 9] +
```

```
        k[g6pdh, 6] k[g6pdh, 9] + k[g6pdh, 7] k[g6pdh, 9]);
D7[g6pdh] = k[g6pdh, 1] k[g6pdh, 4] k[g6pdh, 6] k[g6pdh, 8] ;
D8[g6pdh] = k[g6pdh, 3] k[g6pdh, 5] k[g6pdh, 7] k[g6pdh, 10];
D9[g6pdh] = k[g6pdh, 8] k[g6pdh, 10]
    (k[g6pdh, 2] k[g6pdh, 4] + k[g6pdh, 2] k[g6pdh, 5] +
      k[g6pdh, 2] k[g6pdh, 6] + k[g6pdh, 4] k[g6pdh, 6]);
D10[g6pdh] = k[g6pdh, 1] k[g6pdh, 3] k[g6pdh, 8]
  (k[g6pdh, 5] + k[g6pdh, 6]);
D11[g6pdh] = k[g6pdh, 3] k[g6pdh, 8]
  k[g6pdh, 10] (k[g6pdh, 5] + k[g6pdh, 6]);


v[g6pdh] := G6PDH
    (N1[g6pdh] Glc6P[t] NADP[t] - N2[g6pdh] P6GL[t] NADPH[t]) /
    (D1[g6pdh] + D2[g6pdh] NADP[t] + D3[g6pdh] Glc6P[t] +
      D4[g6pdh] P6GL[t] + D5[g6pdh] NADPH[t] +
      D6[g6pdh] Glc6P[t] NADP[t] + D7[g6pdh] NADP[t] P6GL[t] +
      D8[g6pdh] Glc6P[t] NADPH[t] + D9[g6pdh] P6GL[t] NADPH[t] +
      D10[g6pdh] Glc6P[t] NADP[t] P6GL[t] +
      D11[g6pdh] P6GL[t] NADPH[t] Glc6P[t]);
```

A3.2.2 6-Phosphogluconolactone Hydrolysis

δ-Gluconolactonase

Parameters

```
k[lactonase, 1] = 1.3 * 10^7;
k[lactonase, 2] = 1.0 * 10^3;
k[lactonase, 3] = 2.9 * 10^1;
Lactonase = 14.0 * 10^-6;
```

Rate Equation

```
Km[lactonase, p6gl] =
  (k[lactonase, 2] + k[lactonase, 3]) / k[lactonase, 1];
kcatf[lactonase] = k[lactonase, 3];
```

$$v[\text{lactonase}] := \frac{\text{Lactonase kcatf[lactonase] P6GL}[t]}{Km[\text{lactonase, p6gl}] + \text{P6GL}[t]} \; ;$$

Spontaneous 6-Phosphogluconolactone Hydrolysis

Parameters

```
k[spontaneouspglhydrolysis] = 7.1 * 10^-4;
```

Rate Equation

```
v[spontaneouspglhydrolysis] :=
  +k[spontaneouspglhydrolysis] P6GL[t];
```

Total 6-Phosphogluconolactone Hydrolysis

```
v[pglhydrolysis] := v[lactonase] + v[spontaneouspglhydrolysis];
```

A3.2.3 6-Phosphogluconate Dehydrogenase

Parameters

```
k[p6gdh, 1] = 2 * 1.2 * 10^6;
k[p6gdh, 2] = 4.1 * 10^2;
k[p6gdh, 3] = 2 * 1.0 * 10^9;
k[p6gdh, 4] = 2.6 * 10^4;
k[p6gdh, 5] = 4.8 * 10^1;
k[p6gdh, 6] = 3.0 * 10^1;
k[p6gdh, 7] = 6.3 * 10^2;
k[p6gdh, 8] = 3.6 * 10^4;
k[p6gdh, 9] = 8.0 * 10^2;
k[p6gdh, 10] = 0.5 * 4.5 * 10^5;
k[p6gdh, 11] = 3.0 * 10^2;
k[p6gdh, 12] = 0.5 * 9.9 * 10^6;
P6GDH = 2.1 * 10^-6;
```

Rate Equation

```
N1[p6gdh] = k[p6gdh, 1] k[p6gdh, 3]
  k[p6gdh, 5] k[p6gdh, 7] k[p6gdh, 9] k[p6gdh, 11];
N2[p6gdh] = k[p6gdh, 2] k[p6gdh, 4] k[p6gdh, 6]
  k[p6gdh, 8] k[p6gdh, 10] k[p6gdh, 12];
D1[p6gdh] = k[p6gdh, 2] k[p6gdh, 9]
  k[p6gdh, 11] (k[p6gdh, 4] k[p6gdh, 6] +
    k[p6gdh, 4] k[p6gdh, 7] + k[p6gdh, 5] k[p6gdh, 7]);
D2[p6gdh] = k[p6gdh, 1] k[p6gdh, 9] k[p6gdh, 11]
  (k[p6gdh, 4] k[p6gdh, 6] +
    k[p6gdh, 4] k[p6gdh, 7] + k[p6gdh, 5] k[p6gdh, 7]);
D3[p6gdh] = k[p6gdh, 3] k[p6gdh, 5] k[p6gdh, 7]
  k[p6gdh, 9] k[p6gdh, 11];
D4[p6gdh] = k[p6gdh, 2] k[p6gdh, 4]
  k[p6gdh, 6] k[p6gdh, 8] k[p6gdh, 11];
D5[p6gdh] = k[p6gdh, 2] k[p6gdh, 9]
  k[p6gdh, 12] (k[p6gdh, 4] k[p6gdh, 6] +
    k[p6gdh, 4] k[p6gdh, 7] + k[p6gdh, 5] k[p6gdh, 7]);
D6[p6gdh] = k[p6gdh, 1] k[p6gdh, 3]
```

```
   (k[p6gdh, 5] k[p6gdh, 7] k[p6gdh, 9] + k[p6gdh, 5] k[p6gdh, 7]
      k[p6gdh, 11] + k[p6gdh, 5] k[p6gdh, 9] k[p6gdh, 11] +
      k[p6gdh, 6] k[p6gdh, 9] k[p6gdh, 11] +
      k[p6gdh, 7] k[p6gdh, 9] k[p6gdh, 11]);
D7[p6gdh] = k[p6gdh, 1] k[p6gdh, 4] k[p6gdh, 6]
   k[p6gdh, 8] k[p6gdh, 11];
D8[p6gdh] = k[p6gdh, 3] k[p6gdh, 5]
   k[p6gdh, 7] k[p6gdh, 9] k[p6gdh, 12];
D9[p6gdh] = k[p6gdh, 2] k[p6gdh, 4]
   k[p6gdh, 6] k[p6gdh, 8] k[p6gdh, 10];
D10[p6gdh] = k[p6gdh, 2] k[p6gdh, 4]
   k[p6gdh, 6] k[p6gdh, 8] k[p6gdh, 12];
D11[p6gdh] = k[p6gdh, 2] k[p6gdh, 10]
   k[p6gdh, 12] (k[p6gdh, 4] k[p6gdh, 6] +
      k[p6gdh, 4] k[p6gdh, 7] + k[p6gdh, 5] k[p6gdh, 7]);
D12[p6gdh] = k[p6gdh, 1] k[p6gdh, 3] k[p6gdh, 8]
   k[p6gdh, 11] (k[p6gdh, 5] + k[p6gdh, 6]);
D13[p6gdh] = k[p6gdh, 1] k[p6gdh, 3]
   k[p6gdh, 5] k[p6gdh, 7] k[p6gdh, 10];
D14[p6gdh] = k[p6gdh, 1] k[p6gdh, 4]
   k[p6gdh, 6] k[p6gdh, 8] k[p6gdh, 10];
D15[p6gdh] = k[p6gdh, 3] k[p6gdh, 5]
   k[p6gdh, 7] k[p6gdh, 10] k[p6gdh, 12];
D16[p6gdh] = k[p6gdh, 8] k[p6gdh, 10] k[p6gdh, 12]
   (k[p6gdh, 2] k[p6gdh, 4] + k[p6gdh, 2] k[p6gdh, 5] +
      k[p6gdh, 2] k[p6gdh, 6] + k[p6gdh, 4] k[p6gdh, 6]);
D17[p6gdh] = k[p6gdh, 1] k[p6gdh, 3] k[p6gdh, 8]
   k[p6gdh, 10] (k[p6gdh, 5] + k[p6gdh, 6]);
D18[p6gdh] = k[p6gdh, 3] k[p6gdh, 8] k[p6gdh, 10]
   k[p6gdh, 12] (k[p6gdh, 5] + k[p6gdh, 6]);


v[p6gdh] := P6GDH
   (N1[p6gdh] P6G[t] NADP[t] - N2[p6gdh] CO2[t] Ru5P[t] NADPH[t]) /
   (D1[p6gdh] + D2[p6gdh] NADP[t] + D3[p6gdh] P6G[t] +
      D4[p6gdh] CO2[t] + D5[p6gdh] NADPH[t] +
      D6[p6gdh] P6G[t] NADP[t] + D7[p6gdh] NADP[t] CO2[t] +
      D8[p6gdh] P6G[t] NADPH[t] + D9[p6gdh] CO2[t] Ru5P[t] +
      D10[p6gdh] CO2[t] NADPH[t] + D11[p6gdh] Ru5P[t] NADPH[t] +
      D12[p6gdh] P6G[t] NADP[t] CO2[t] +
      D13[p6gdh] P6G[t] NADP[t] Ru5P[t] +
      D14[p6gdh] NADP[t] CO2[t] Ru5P[t] + D15[p6gdh] P6G[t]
       Ru5P[t] NADPH[t] + D16[p6gdh] CO2[t] Ru5P[t] NADPH[t] +
      D17[p6gdh] P6G[t] NADP[t] CO2[t] Ru5P[t] +
      D18[p6gdh] P6G[t] CO2[t] Ru5P[t] NADPH[t]);
```

A3.2.4 Glutathione Reductase

Parameters

```
k[gssgr, 1] = 8.5 * 10^7;
k[gssgr, 2] = 5.1 * 10^2;
k[gssgr, 3] = 1.0 * 10^9;
k[gssgr, 4] = 7.2 * 10^4;
k[gssgr, 5] = 8.1 * 10^2;
k[gssgr, 6] = 1.0 * 10^3;
k[gssgr, 7] = 1.0 * 10^6;
k[gssgr, 8] = 5.0 * 10^7;
k[gssgr, 9] = 1.0 * 10^6;
k[gssgr, 10] = 5.0 * 10^7;
k[gssgr, 11] = 7.0 * 10^3;
k[gssgr, 12] = 1.0 * 10^8;
GSSGR = 1.25 * 10^-7;
```

Rate Equation

```
N1[gssgr] = k[gssgr, 1] k[gssgr, 3]
   k[gssgr, 5] k[gssgr, 7] k[gssgr, 9] k[gssgr, 11];
N2[gssgr] = k[gssgr, 2] k[gssgr, 4] k[gssgr, 6]
  k[gssgr, 8] k[gssgr, 10] k[gssgr, 12]; D1[gssgr] =
 k[gssgr, 2] (k[gssgr, 4] k[gssgr, 6] + k[gssgr, 4] k[gssgr, 7] +
   k[gssgr, 5] k[gssgr, 7]) k[gssgr, 9] k[gssgr, 11];
D2[gssgr] = k[gssgr, 1] (k[gssgr, 4] k[gssgr, 6] +
    k[gssgr, 4] k[gssgr, 7] + k[gssgr, 5] k[gssgr, 7])
   k[gssgr, 9] k[gssgr, 11];
D3[gssgr] = k[gssgr, 3] k[gssgr, 5] k[gssgr, 7]
   k[gssgr, 9] k[gssgr, 11];
D4[gssgr] = k[gssgr, 2] k[gssgr, 4] k[gssgr, 6]
   k[gssgr, 8] k[gssgr, 11];
D5[gssgr] = k[gssgr, 2]
   (k[gssgr, 4] k[gssgr, 6] + k[gssgr, 4] k[gssgr, 7] +
    k[gssgr, 5] k[gssgr, 7]) k[gssgr, 9] k[gssgr, 12];
D6[gssgr] = k[gssgr, 1] k[gssgr, 3]
   (k[gssgr, 5] k[gssgr, 7] k[gssgr, 9] + k[gssgr, 5] k[gssgr, 7]
     k[gssgr, 11] + k[gssgr, 5] k[gssgr, 9] k[gssgr, 11] +
    k[gssgr, 6] k[gssgr, 9] k[gssgr, 11] +
    k[gssgr, 7] k[gssgr, 9] k[gssgr, 11]);
D7[gssgr] = k[gssgr, 1] k[gssgr, 4] k[gssgr, 6]
   k[gssgr, 8] k[gssgr, 11];
D8[gssgr] = k[gssgr, 3] k[gssgr, 5] k[gssgr, 7]
   k[gssgr, 9] k[gssgr, 12];
```

```
D9[gssgr] = k[gssgr, 2] k[gssgr, 4] k[gssgr, 6]
   k[gssgr, 8] k[gssgr, 10];
D10[gssgr] = k[gssgr, 2] k[gssgr, 4]
   k[gssgr, 6] k[gssgr, 8] k[gssgr, 12];
D11[gssgr] = k[gssgr, 2]
   (k[gssgr, 4] k[gssgr, 6] + k[gssgr, 4] k[gssgr, 7] +
     k[gssgr, 5] k[gssgr, 7]) k[gssgr, 10] k[gssgr, 12];
D12[gssgr] = k[gssgr, 1] k[gssgr, 3] (k[gssgr, 5] + k[gssgr, 6])
   k[gssgr, 8] k[gssgr, 11]; D13[gssgr] =
  k[gssgr, 1] k[gssgr, 3] k[gssgr, 5] k[gssgr, 7] k[gssgr, 10];
D14[gssgr] = k[gssgr, 1] k[gssgr, 4]
   k[gssgr, 6] k[gssgr, 8] k[gssgr, 10];
D15[gssgr] = k[gssgr, 3] k[gssgr, 5] k[gssgr, 7]
   k[gssgr, 10] k[gssgr, 12];
D16[gssgr] = (k[gssgr, 2] k[gssgr, 4] + k[gssgr, 2] k[gssgr, 5] +
     k[gssgr, 2] k[gssgr, 6] + k[gssgr, 4] k[gssgr, 6])
   k[gssgr, 8] k[gssgr, 10] k[gssgr, 12];
D17[gssgr] = k[gssgr, 1] k[gssgr, 3]
   (k[gssgr, 5] + k[gssgr, 6]) k[gssgr, 8] k[gssgr, 10];
D18[gssgr] = k[gssgr, 3] (k[gssgr, 5] + k[gssgr, 6])
   k[gssgr, 8] k[gssgr, 10] k[gssgr, 12];


v[gssgr] := (GSSGR
      (N1[gssgr] NADPH[t] GSSG[t] - N2[gssgr] GSH[t]^2 NADP[t] )) /
      (D1[gssgr] + D4[gssgr] GSH[t] + D9[gssgr] GSH[t]^2 +
       D3[gssgr] GSSG[t] + D5[gssgr] NADP[t] +
       D10[gssgr] GSH[t] NADP[t] + D11[gssgr] GSH[t] NADP[t] +
       D16[gssgr] GSH[t]^2 NADP[t] + D8[gssgr] GSSG[t] NADP[t] +
       D15[gssgr] GSH[t] GSSG[t] NADP[t] +
       D18[gssgr] GSH[t]^2 GSSG[t] NADP[t] +
       D2[gssgr] NADPH[t] + D7[gssgr] GSH[t] NADPH[t] +
       D14[gssgr] GSH[t]^2 NADPH[t] + D6[gssgr] GSSG[t] NADPH[t] +
       D12[gssgr] GSH[t] GSSG[t] NADPH[t] + D13[gssgr] GSH[t]
        GSSG[t] NADPH[t] + D17[gssgr] GSH[t]^2 GSSG[t] NADPH[t]);
```

A3.2.5 Ribulose-5-Phosphate Epimerase

Parameters

```
k[ru5pe, 1] = 3.91 * 10^6;
k[ru5pe, 2] = 4.38 * 10^2;
k[ru5pe, 3] = 3.05 * 10^2;
k[ru5pe, 4] = 1.49 * 10^6;
Ru5PE = 4.22 * 10^-6;
```

Rate Equation

```
Km[ru5pe, ru5p] = (k[ru5pe, 2] + k[ru5pe, 3]) / k[ru5pe, 1];
Km[ru5pe, xu5p] = (k[ru5pe, 2] + k[ru5pe, 3]) / k[ru5pe, 4];
kcatf[ru5pe] = k[ru5pe, 3];
kcatr[ru5pe] = k[ru5pe, 2];
```

$$\text{ru5perd} := 1 + \frac{\text{Ru5P[t]}}{\text{Km[ru5pe, ru5p]}} + \frac{\text{Xu5P[t]}}{\text{Km[ru5pe, xu5p]}} ;$$

$$v[\text{ru5pe}] :=$$
$$\frac{\text{Ru5PE}}{\text{ru5perd}} \left(\frac{\text{kcatf[ru5pe] Ru5P[t]}}{\text{Km[ru5pe, ru5p]}} - \frac{\text{kcatr[ru5pe] Xu5P[t]}}{\text{Km[ru5pe, xu5p]}} \right) ;$$

A3.2.6 Ribose-5-Phosphate Isomerase

Parameters

```
k[r5pi, 1] = 6.09 * 10^4;
k[r5pi, 2] = 3.33 * 10^1;
k[r5pi, 3] = 1.42 * 10^1;
k[r5pi, 4] = 2.16 * 10^4;
R5PI = 1.42 * 10^-5;
```

Rate Equation

```
Km[r5pi, ru5p] = (k[r5pi, 2] + k[r5pi, 3]) / k[r5pi, 1];
Km[r5pi, rib5p] = (k[r5pi, 2] + k[r5pi, 3]) / k[r5pi, 4];
kcatf[r5pi] = k[r5pi, 3];
kcatr[r5pi] = k[r5pi, 2];
```

$$\text{r5pird} := 1 + \frac{\text{Ru5P[t]}}{\text{Km[r5pi, ru5p]}} + \frac{\text{Xu5P[t]}}{\text{Km[r5pi, rib5p]}} ;$$

$$v[\text{r5pi}] :=$$
$$\frac{\text{R5PI}}{\text{r5pird}} \left(\frac{\text{kcatf[r5pi] Ru5P[t]}}{\text{Km[r5pi, ru5p]}} - \frac{\text{kcatr[r5pi] Rib5P[t]}}{\text{Km[r5pi, rib5p]}} \right) ;$$

A3.2.7 Transketolase

Parameters

```
k[tk, 1] = 2.16 * 10^5;
k[tk, 2] = 3.8 * 10^1;
k[tk, 3] = 3.4 * 10^1;
k[tk, 4] = 1.56 * 10^5;
k[tk, 5] = 3.29 * 10^5;
```

```
k[tk, 6] = 1.75 * 10 ^ 2;
k[tk, 7] = 4.0 * 10 ^ 1;
k[tk, 8] = 4.48 * 10 ^ 4;
k[tk, 9] = 2.24 * 10 ^ 6;
k[tk, 10] = 1.75 * 10 ^ 2;
k[tk, 11] = 4.0 * 10 ^ 1;
k[tk, 12] = 2.13 * 10 ^ 4;
```

Rate Equations

```
v[tk1] := +k[tk, 1] * TK[t] * Xu5P[t] - k[tk, 2] * TK$Xu5P[t] ;
v[tk2] := +k[tk, 3] * TK$Xu5P[t] - k[tk, 4] * TKG[t] * GraP[t] ;
v[tk3] := +k[tk, 5] * TKG[t] * Rib5P[t] - k[tk, 6] * TKG$Rib5P[t];
v[tk4] := +k[tk, 7] * TKG$Rib5P[t] - k[tk, 8] * TK[t] * Sed7P[t];
v[tk5] := +k[tk, 9] * TKG[t] * Ery4P[t] - k[tk, 10] * TKG$Ery4P[t];
v[tk6] := +k[tk, 11] * TKG$Ery4P[t] - k[tk, 12] * TK[t] * Fru6P[t];
```

A3.2.8 Transaldolase

Parameters

```
k[ta, 1] = 5.8 * 10 ^ 5;
k[ta, 2] = 4.53 * 10 ^ 1;
k[ta, 3] = 1.63 * 10 ^ 1;
k[ta, 4] = 1.01 * 10 ^ 6;
k[ta, 5] = 4.9 * 10 ^ 5;
k[ta, 6] = 6.0 * 10 ^ 1;
k[ta, 7] = 1.7 * 10 ^ 1;
k[ta, 8] = 7.9 * 10 ^ 4;
TA = 0.69 * 10 ^ -6;
```

Rate Equations

```
N1[ta] = k[ta, 1] k[ta, 3] k[ta, 5] k[ta, 7];
N2[ta] = k[ta, 2] k[ta, 4] k[ta, 6] k[ta, 8];
D1[ta] = k[ta, 1] k[ta, 3] (k[ta, 6] + k[ta, 7]);
D2[ta] = k[ta, 5] k[ta, 7] ( k[ta, 2] + k[ta, 3]);
D3[ta] = k[ta, 2] k[ta, 4] (k[ta, 6] + k[ta, 7]);
D4[ta] = k[ta, 6] k[ta, 8] ( k[ta, 2] + k[ta, 3]);
D5[ta] = k[ta, 1] k[ta, 5] (k[ta, 3] + k[ta, 7]);
D6[ta] = k[ta, 4] k[ta, 8] (k[ta, 2] + k[ta, 6]);
D7[ta] = k[ta, 5] k[ta, 8] ( k[ta, 2] + k[ta, 3]);
D8[ta] = k[ta, 1] k[ta, 4] (k[ta, 6] + k[ta, 7]);

v[ta] := TA (N1[ta] Sed7P[t] GraP[t] - N2[ta] Ery4P[t] Fru6P[t]) /
    (D1[ta] Sed7P[t] + D2[ta] GraP[t] +
```

```
        D3[ta] Ery4P[t] + D4[ta] Fru6P[t] +
        D5[ta] Sed7P[t] GraP[t] + D6[ta] Ery4P[t] Fru6P[t] +
        D7[ta] GraP[t] Fru6P[t] + D8[ta] Sed7P[t] Ery4P[t]);
```

A3.3 2,3-BPG Shunt Enzyme

A3.3.1 2,3-Bisphosphoglycerate Synthase-Phosphatase

Parameters

$$k[bpgsp, 1] := 0.8 * 10\wedge 8 \left(\frac{1 + 10\wedge-6.8 / 10\wedge-7.20}{1 + 10\wedge-6.8 / 10\wedge-pH1[t]} \right);$$

$$k[bpgsp, 2] := 4.0 * 10\wedge 2;$$

$$k[bpgsp, 3] := 9.9 \left(\frac{1 + (10\wedge-7.2 / 10\wedge-7.17)\wedge 4}{1 + (10\wedge-pH1[t] / 10\wedge-7.17)\wedge 4} \right);$$

$$k[bpgsp, 4] := 1.85 * 10\wedge 8 \left(\frac{1 + (10\wedge-7.2 / 10\wedge-7.17)\wedge 4}{1 + (10\wedge-pH1[t] / 10\wedge-7.17)\wedge 4} \right);$$

$$k[bpgsp, 5] = 1.0 * 10\wedge 8;$$

$$k[bpgsp, 6] := 1 * 10\wedge 3 \left(\frac{1 + (10\wedge-7.2 / 10\wedge-7.17)\wedge 4}{1 + (10\wedge-pH1[t] / 10\wedge-7.17)\wedge 4} \right);$$

$$k[bpgsp, 7] = 1 * 10\wedge 3;$$

$$k[bpgsp, 8] = 1 * 10\wedge 4;$$

$$k[bpgsp, 9] = 0.55;$$

$$k[bpgsp, 10] = 1.979 * 10\wedge 3;$$

$$k[bpgsp, 11] = 1 * 10\wedge-2;$$

$$k[bpgsp, 12] = 1 * 10\wedge 3;$$

$$k[bpgsp, 13] := 1.8 * 10\wedge 6 \left(\frac{1 + 10\wedge-6.8 / 10\wedge-7.20}{1 + 10\wedge-6.8 / 10\wedge-pH1[t]} \right);$$

$$k[bpgsp, 14] = 1.0 * 10\wedge 9;$$

$$k[bpgsp, 15] = 6.1 * 10\wedge 5;$$

$$k[bpgsp, 16] = 0.19;$$

Rate Equations

```
v[bpgsp1] :=
    +k[bpgsp, 1] * BPGSP[t] B13PG[t] - k[bpgsp, 2] * BPGSP$B13PG[t];
v[bpgsp2] := +k[bpgsp, 3] * BPGSP$B13PG[t];
v[bpgsp3] := +k[bpgsp, 4] * BPGSPP[t] * P3GA[t] -
    k[bpgsp, 5] * BPGSPP$P3GA[t];
v[bpgsp4] := +k[bpgsp, 6] * BPGSPP[t] * P2GA[t] -
    k[bpgsp, 7] * BPGSPP$P2GA[t];
v[bpgsp5] := +k[bpgsp, 8] * BPGSPP$P3GA[t] -
    k[bpgsp, 9] * BPGSP$B23PG[t];
v[bpgsp6] := +k[bpgsp, 10] * BPGSPP$P2GA[t] -
    k[bpgsp, 11] * BPGSP$B23PG[t];
```

```
v[bpgsp7] := +k[bpgsp, 12] * BPGSP$B23PG[t] -
    k[bpgsp, 13] * BPGSP[t] * B23PG[t];
v[bpgsp8] := +k[bpgsp, 14] * BPGSPP[t] * Phos[t] -
    k[bpgsp, 15] * BPGSPP$Phos[t];
v[bpgsp9] := k[bpgsp, 16] * BPGSPP$Phos[t];
```

A3.4 Energy Consumption and Oxidative Reactions

A3.4.1 Adenylate Kinase

Parameters

```
K[hamp] = 3.09 * 10^6;
K[kamp] = 1.8;
k[ak, 1] =
    4.3 * 10^3 (1 + 10^-pH1[t] * K[hadp] + k[+1] * K[kadp]);
k[ak, 2] = 1.4 * 10^3 (1 + 10^-pH1[t] * K[hamp] + k[+1] * K[kamp]);
```

Rate Equation

```
v[ak] := +k[ak, 1] * MgADP[t] * ADP[t] - k[ak, 2] MgATP[t] AMP[t];
```

A3.4.2 ATP Consuming Processes

Parameters

```
k[atpase] := 5.85 * 10^-4;
```

Rate Equation

```
v[atpase] := +k[atpase] MgATP[t];
```

A3.4.3 Glutathione Oxidation

Parameters

```
k[ox] := 3.4 * 10^-5;
```

Rate Equation

```
v[ox] := k[ox] GSH[t];
```

A3.4.4 Non-Glycolytic NADH Consumption

Parameters

```
k[oxNADH] := 16.3 * 10^-3;
```

Rate Equation

```
v[oxnadh] := k[oxNADH] * NADH[t];
```

A3.5 Membrane Transport

A3.5.1 Pyruvate Transport

Parameters

```
k[pyrtransport, i] = 1.8 * 10^-2;
k[pyrtransport, o] := k[pyrtransport, i] / r[t] * Vol_e / Vol_i;
```

Rate Equation

```
v[pyrtransport] := +k[pyrtransport, o] * Pyr[t] * Vol_i -
    k[pyrtransport, i] * Pyre[t] * Vol_e;
```

A3.5.2 Lactate Transport

Parameters

```
k[lactransport, i] = 3.6 * 10^-3;
Keq[lactransport] :=
  (1 + 10^(pH1[t] - 3.73)) / (1 + 10^(pH1[t] - 3.73) / r[t]);
k[lactransport, o] :=
  k[lactransport, i] / Keq[lactransport] * Vol_e / Vol_i;
```

Rate Equation

```
v[lactransport] := +k[lactransport, o] * Lac[t] * Vol_i -
    k[lactransport, i] * Lace[t] * Vol_e;
```

A3.5.3 Phosphate Transport

Parameters

```
k[phostransport, i] = 5.6 * 10^-4;
k[phostransport, o] :=
  k[phostransport, i] / Keq[phostransport] * Vol_e / Vol_i;
Keq[phostransport] := (1 + 10^(pH1[t] - 6.75)) /
    (1 / r[t] + 10^(pH1[t] - 6.75) / r[t]^2);
```

Rate Equation

```
v[phostransport] := +k[phostransport, o] * Phos[t] * Vol_i -
    k[phostransport, i] * Phose[t] * Vol_e;
```

A3.6 Magnesium-Metabolite Binding

A3.6.1 MgATP

Parameters

```
K[mgatp] = 4.32 * 10 ^ 4;
K[hatp] = 9.07 * 10 ^ 6;
K[mghatp] = 7.48 * 10 ^ 2;
K[katp] = 14;
k[mgatp, a] := 3.12 * 10 ^ 7 * 8.4 * 10 ^ -5
    ( K[mgatp] + 10 ^ -pH1[t] K[hatp] K[mghatp]
    ----------------------------------------------- );
      1 + 10 ^ -pH1[t] K[hatp] + k[+1] K[katp]
k[mgatp, d] = 1.2 * 10 ^ 3;
```

Rate Equation

```
v[mgatp] := +k[mgatp, a] * ATP[t] * Mg[t] - k[mgatp, d] * MgATP[t];
```

A3.6.2 MgADP

Parameters

```
K[mgadp] = 3.29 * 10 ^ 3;
K[hadp] = 5.42 * 10 ^ 6;
K[mghadp] = 1.07 * 10 ^ 2;
K[kadp] = 4.8;
k[mgadp, a] := 2.76 * 10 ^ 6 * 6.2 * 10 ^ -4
    ( K[mgadp] + 10 ^ -pH1[t] K[hadp] K[mghadp]
    ----------------------------------------------- );
      1 + 10 ^ -pH1[t] K[hadp] + k[+1] K[kadp]
k[mgadp, d] = 1.2 * 10 ^ 3;
```

Rate Equation

```
v[mgadp] := +k[mgadp, a] * ADP[t] * Mg[t] - k[mgadp, d] * MgADP[t];
```

A3.6.3 Mg2,3-BPG

Parameters

```
K[mgbpg] = 7.41 * 10 ^ 3;
K[hbpg] = 1.62 * 10 ^ 8;
K[mghbpg] = 5.13 * 10 ^ 2;
K[h2bpg] = 4.27 * 10 ^ 6;
K[kbpg] = 85.1;
K[khbpg] = 8.9;
mgbpgphf :=
```

```
    3.2 * 10 ^ -3 (K[mgbpg] + 10 ^ -pH1[t] K[hbpg] K[mghbpg]) /
        (1 + 10 ^ -pH1[t] K[hbpg] + 10 ^ - (2 * pH1[t]) K[hbpg] K[h2bpg] +
            k[+1] K[kbpg] + k[+1] * 10 ^ -pH1[t] K[hbpg] K[khbpg]);
    k[mgbpg, a] := 8.04 * 10 ^ 5 * mgbpgphf;
    k[mgbpg, d] = 1.2 * 10 ^ 3;
```

Rate Equation

```
    v[mgb23pg] :=
        +k[mgbpg, a] * B23PG[t] * Mg[t] - k[mgbpg, d] * Mg$B23PG[t];
```

A3.6.4 Mg1,3-BPG

Parameters

```
    k[mgb13pg, a] := 2.28 * 10 ^ 5 * mgbpgphf;
    k[mgb13pg, d] = 1.2 * 10 ^ 3;
```

Rate Equation

```
    v[mgb13pg] :=
        +k[mgb13pg, a] * B13PG[t] * Mg[t] - k[mgb13pg, d] * Mg$B13PG[t];
```

A3.6.5 MgFructose (1,6)-Bisphosphate

Parameters

```
    K[mgfru16p2] = 3.63 * 10 ^ 2;
    K[hf] = 7.56 * 10 ^ 6;
    K[mghf] = 8.9 * 10 ^ 1;
    K[h2f] = 1.12 * 10 ^ 6;
    K[kf] = 10.7;
    K[khf] = 3.3;
    mgfphf :=
        8.3 * 10 ^ -3 * ((K[mgfru16p2] + 10 ^ -pH1[t] K[hf] K[mghf]) /
            (1 + 10 ^ -pH1[t] K[hf] + 10 ^ - (2 * pH1[t]) K[hf] K[h2f] +
                k[+1] K[kf] + k[+1] * 10 ^ -pH1[t] K[hf] K[khf]));
    k[mgf16p2, a] := 4.80 * 10 ^ 5 * mgfphf;
    k[mgf16p2, d] = 1.2 * 10 ^ 3;
```

Rate Equation

```
    v[mgfru16p2] := +k[mgf16p2, a] * Fru16P2[t] * Mg[t] -
        k[mgf16p2, d] * Mg$Fru16P2[t];
```

A3.6.6 MgGlucose (1,6)-Bisphosphate

Rate Equation

```
v[mgglc16p2] := +k[mgf16p2, a] * Glc16P2[t] * Mg[t] -
   k[mgf16p2, d] * Mg$Glc16P2[t];
```

A3.6.7 MgPhosphate

Parameters

```
K[hphos] = 5.68 * 10^6;
K[kphos] = 3.0;
k[mgphos, a] :=
```
$$k[mgphos, a] := 4.08 * 10^4 * \frac{1 + 10^{-7.2} K[hphos] + 0.15 K[kphos]}{1 + 10^{-pH1[t]} K[hphos] + k[+1] K[kphos]};$$
```
k[mgphos, d] = 1.2 * 10^3;
```

Rate Equation

```
v[mgphos] :=
   +k[mgphos, a] * Phos[t] * Mg[t] - k[mgphos, d] * Mg$Phos[t];
```

A3.7 pH Depedendence of Hb-Metabolite Binding

```
Ka[hb] = 10^-6.6;
```

$$hbphf := \frac{1 + (2 \, Ka[hb] / 10^{-7.2}) + (Ka[hb] / 10^{-7.2})^2}{1 + (2 \, Ka[hb] / 10^{-pH1[t]}) + (Ka[hb] / 10^{-pH1[t]})^2};$$

A3.8 Oxy-Haemoglobin-Metabolite Binding

A3.8.1 Hb-MgATP

Parameters

```
k[hbmgatp, a] := 4.68 * 10^4 * hbphf;
k[hbmgatp, d] = 1.2 * 10^3;
```

Rate Equation

```
v[hbmgatp] :=
   +k[hbmgatp, a] * Hb[t] * MgATP[t] - k[hbmgatp, d] * Hb$MgATP[t];
```

A3.8.2 Hb-ATP

Parameters

```
k[hbatp, a] := 4.32 * 10 ^ 5 * hbphf;
k[hbatp, d] = 1.2 * 10 ^ 3;
```

Rate Equation

```
v[hbatp] := +k[hbatp, a] * ATP[t] * Hb[t] - k[hbatp, d] * Hb$ATP[t];
```

A3.8.3 Hb-ADP

Parameters

```
k[hbadp, a] := 3.0 * 10 ^ 5 * hbphf;
k[hbadp, d] = 1.2 * 10 ^ 3;
```

Rate Equation

```
v[hbadp] := +k[hbadp, a] * ADP[t] * Hb[t] - k[hbadp, d] * Hb$ADP[t];
```

A3.8.4 Hb-23BPG

Parameters

```
k[hbbpg, a] := 3.0 * 10 ^ 5 * hbphf;
k[hbbpg, d] = 1.2 * 10 ^ 3;
```

Rate Equation

```
v[hbbpg] :=
   +k[hbbpg, a] * Hb[t] * B23PG[t] - k[hbbpg, d] * Hb$B23PG[t];
```

A3.8.5 Hb-13BPG

Parameters

```
k[hbb13pg, a] := 3.80 * 10 ^ 5 * hbphf;
k[hbb13pg, d] = 1.2 * 10 ^ 3;
```

Rate Equation

```
v[hbb13pg] :=
 +k[hbb13pg, a] * Hb[t] * B13PG[t] - k[hbb13pg, d] * Hb$B13PG[t]
```

A3.9 Deoxy-Haemoglobin-Metabolite Binding

A3.9.1 deoxyHb-MgATP

Parameters

```
k[hbdmgatp, a] := 1.68 * 10^5 * hbphf;
k[hbdmgatp, d] = 1.2 * 10^3;

v[hbdmgatp] := +k[hbdmgatp, a] * Hbd[t] * MgATP[t] -
    k[hbdmgatp, d] * Hbd$MgATP[t];
```

A3.9.2 deoxyHb-ATP

Parameters

```
k[hbdatp, a] := 3.12 * 10^6 * hbphf;
k[hbdatp, d] = 1.2 * 10^3;
```

Rate Equation

```
v[hbdatp] :=
    +k[hbdatp, a] * ATP[t] * Hbd[t] - k[hbdatp, d] * Hbd$ATP[t];
```

A3.9.3 deoxyHb-ADP

Parameters

```
k[hbdadp, a] := 1.44 * 10^6 * hbphf;
k[hbdadp, d] = 1.2 * 10^3;
```

Rate Equation

```
v[hbdadp] :=
    +k[hbdadp, a] * ADP[t] * Hbd[t] - k[hbdadp, d] * Hbd$ADP[t];
```

A3.9.4 deoxyHb-23BPG

Parameters

```
k[hbdbpg, a] := 6.00 * 10^6 * hbphf;
k[hbdbpg, d] = 1.2 * 10^3;
```

Rate Equation

```
v[hbdbpg] :=
    +k[hbdbpg, a] * Hbd[t] * B23PG[t] - k[hbdbpg, d] * Hbd$B23PG[t];
```

A3.9.5 deoxyHb-Fructose (1,6)-bisphosphate

Parameters

```
k[hbdfbp, a] := 1.21 * 10 ^ 6 * hbphf;
k[hbdfbp, d] = 1.2 * 10 ^ 3;
```

Rate Equation

```
v[hbdfbp] := +k[hbdfbp, a] * Hbd[t] * Fru16P2[t] -
    k[hbdfbp, d] * Hbd$Fru16P2[t];
```

A3.9.6 deoxyHb-13BPG

Parameters

```
k[hbdb13pg, a] := 1.86 * 10 ^ 6 * hbphf;
k[hbdb13pg, d] = 1.2 * 10 ^ 3;
```

Rate Equation

```
v[hbdb13pg] := +k[hbdb13pg, a] * Hbd[t] * B13PG[t] -
    k[hbdb13pg, d] * Hbd$B13PG[t];

DumpSave["RBCequations"]
```

Appendix 4 - Initial Conditions and External Parameters for the Erythrocyte Model

This appendix contains the external parameters and the initial conditions for the red blood cell model presented in Chapter 7. This appendix should be evaluated so that its contents can be easily loaded using the command **<<initialconditions'**.

A4.1 External Parameters

Define intra- and extracellular volumes.

A4.1.1 Cell Water Fraction of Total Cell Volume

$$\alpha = \frac{7}{10} \, ;$$

A4.1.2 Hematocrit

$$Ht = \frac{1}{2} \, ;$$

A4.1.3 Extracellular Volume

$$Vol_e = 1 - Ht \, ;$$

A4.1.4 Intracellular Volume

$$Vol_i = \alpha \, Ht \, ;$$

A4.1.5 Donnan Ratio

$$r[t] = 0.69 \, ;$$

A4.1.6 Intracelluar K^+ Concentration

```
k[+1] = 0.15;
```

A4.1.7 Intracellular pH

```
pH1[t] = 7.2 ;
```

A4.1.8 Intracellular Carbon Dioxide Concentration

```
CO2[t] = 1.2*10^-3 ;
```

A4.1.9 Other Metabolites (See Figure 7.1 for Abbreviations)

```
CO2[t] = 1.2 × 10⁻³;
Glc[t] = 5 × 10⁻³ ;
Lace[t] = 1.82 × 10⁻³ ;
Phose[t] = 1.92 × 10⁻³ ;
Pyre[t] = 85 × 10⁻⁶ ;
```

A4.2 Initial conditions

See Figure 7.1 for definition of abbreviations.

```
ic1 = {
  ADP[0] == 0.31 × 10⁻³,
  AMP[0] == 30 × 10⁻⁶,
  ATP[0] == 2.1 × 10⁻³,
  B13PG[0] == 0.7 × 10⁻⁶,
  B23PG[0] == 6.70 × 10⁻³,
  BPGSP[0] == 3.8 × 10⁻⁶,
  BPGSPP[0] == 0,
  BPGSPP$P2GA[0] == 0,
  BPGSPP$P3GA[0] == 0,
  BPGSPP$Phos[0] == 0,
  BPGSP$B13PG[0] == 0,
  BPGSP$B23PG[0] == 0,
  Ery4P[0] == 10 × 10⁻⁶,
  Fru16P2[0] == 2.7 × 10⁻⁶,
  Fru6P[0] == 13 × 10⁻⁶,
  Glc16P2[0] == 122 × 10⁻⁶,
  Glc6P[0] == 40 × 10⁻⁶,
  GraP[0] == 5.7 × 10⁻⁶,
```

```
GrnP[0] == 19.0 × 10^-6,
GSH[0] == 3.2 × 10^-3,
GSSG[0] == 0.09 × 10^-6,
Hb[0] == 7 × 10^-3,
Hb$ADP[0] == 0,
Hb$ATP[0] == 0,
Hb$B13PG[0] == 0,
Hb$B23PG[0] == 0,
Hb$MgATP[0] == 0,
Lac[0] == 1.4 * 10^-3,
Mg[0] == 3.0 * 10^-3,
MgADP[0] == 0,
MgATP[0] == 0,
Mg$B13PG[0] == 0,
Mg$B23PG[0] == 0,
Mg$Fru16P2[0] == 0,
Mg$Glc16P2[0] == 0,
Mg$Phos[0] == 0,
NAD[0] == 60 × 10^-6,
NADH[0] == 0.14 × 10^-6,
NADP[0] == 0.125 × 10^-6,
NADPH[0] == 64 × 10^-6,
P2GA[0] == 10 × 10^-6,
P3GA[0] == 64 × 10^-6,
P6G[0] == 1.4 × 10^-7,
P6GL[0] == 1.4 × 10^-10,
PEP[0] == 23 × 10^-6,
Phos[0] == 1.0 × 10^-3,
Pyr[0] == 60 × 10^-6,
Rib5P[0] == 10 × 10^-6,
Ru5P[0] == 10 × 10^-6,
Sed7P[0] == 10 × 10^-6,
TK[0] == 3.3 × 10^-7,
TKG[0] == 0,
TKG$Ery4P[0] == 0,
TKG$Rib5P[0] == 0,
TK$Xu5P[0] == 0,
Xu5P[0] == 1 × 10^-6};

DumpSave["initialconditions"];
```

Appendix 5 - Equation List Describing the Erythrocyte Model of Chapters 7 and 8

This appendix contains the reaction list for the red blood cell model described in Chapter 7. For the definition of each reaction and metabolite label, see Fig. 7.1. This appendix should be evaluated so that its contents can be easily loaded using the command **<<eqns'**.

```
eqns = {
    (*Glycolytic reactions.*)
    {hk,    Glc[t]      +  MgATP[t]   →  Glc6P[t]  +  MgADP[t]  },
            ─────────      ────────      ────────     ────────
            Vol_i + Vol_e    Vol_i         Vol_i       Vol_i

    {gpi,   Glc6P[t]  →  Fru6P[t]} ,
    {pfk,   Fru6P[t] + MgATP[t]  →  Fru16P2[t] + MgADP[t]},
    {ald,   Fru16P2[t]  →  GrnP[t]  + GraP[t]},
    {tpi,   GraP[t] → GrnP[t] },
    {gapdh, GraP[t] + Phos[t] + NAD[t] → B13PG[t] + NADH[t]},
    {pgk,   B13PG[t] + MgADP[t] → P3GA[t] + MgATP[t]},
    {pgm,   P3GA[t] → P2GA[t]},
    {eno,   P2GA[t]  → PEP[t]},
    {pk,    PEP[t] + MgADP[t] → Pyr[t]  + MgATP[t]},
    {ldh,   Pyr[t] + NADH[t] → Lac[t] + NAD[t] },
    {ldhp,  Pyr[t] + NADPH[t] → Lac[t] + NADP[t]},


    (*Reactions of 2,3 BPG synthase-phosphatase.*)
    {bpgsp1,  B13PG[t]  + BPGSP[t]  ↔  BPGSP$B13PG[t] },
    {bpgsp2,  BPGSP$B13PG[t] →  BPGSPP[t] + P3GA[t]},
    {bpgsp3,  BPGSPP[t] + P3GA[t]  →  BPGSPP$P3GA[t] },
    {bpgsp4,  BPGSPP[t] + P2GA[t]  → BPGSPP$P2GA[t]},
    {bpgsp5,  BPGSPP$P3GA[t]   →  BPGSP$B23PG[t] },
    {bpgsp6,  BPGSPP$P2GA[t]  →  BPGSP$B23PG[t] },
    {bpgsp7,  BPGSP$B23PG[t]  → BPGSP[t] + B23PG[t] },
```

```
{bpgsp8,  BPGSPP[t] + Phos[t] → BPGSPP$Phos[t] },
{bpgsp9,  BPGSPP$Phos[t] → BPGSP[t] + 2 Phos[t]},
```

(*Pentose phosphate pathway reactions.*)
```
{g6pdh,         Glc6P[t] + NADP[t] → P6GL[t] + NADPH[t] },
{pglhydrolysis, P6GL[t] → P6G[t] },
{p6gdh,
 P6G[t] + NADP[t] → CO2[t] + Ru5P[t] + NADPH[t]},
{gssgr,         GSSG[t] + NADPH[t] → 2 GSH[t] + NADP[t] },
{ru5pe,         Ru5P[t] → Xu5P[t]},
{r5pi,          Ru5P[t] → Rib5P[t]},
{tk1,           TK[t] + Xu5P[t] → TK$Xu5P[t]},
{tk2,           TK$Xu5P[t] → TKG[t] + GraP[t]},
{tk3,           TKG[t] + Rib5P[t] → TKG$Rib5P[t]},
{tk4,           TKG$Rib5P[t] → TK[t] + Sed7P[t]},
{tk5,           TKG[t] + Ery4P[t] → TKG$Ery4P[t] },
{tk6,           TKG$Ery4P[t] → TK[t] + Fru6P[t] },
{ta,            Sed7P[t] + GraP[t] → Ery4P[t] + Fru6P[t]},
```

(*Energy consumption and oxidative reactions.*)
```
{ak,     MgADP[t] + ADP[t] → MgATP[t] + AMP[t]},
{atpase, MgATP[t] → MgADP[t] + Phos[t]},
{ox,     2 GSH[t] → GSSG[t] },
{oxnadh, NADH[t] → NAD[t]},
```

(*Membrane transport.*)
$$\left\{\text{lactransport, } \frac{1}{\text{Vol}_i}\,\text{Lac}[t] \to \frac{1}{\text{Vol}_e}\,\text{Lace}[t]\right\},$$
$$\left\{\text{pyrtransport, } \frac{1}{\text{Vol}_i}\,\text{Pyr}[t] \to \frac{1}{\text{Vol}_e}\,\text{Pyre}[t]\right\},$$
$$\left\{\text{phostransport, } \frac{1}{\text{Vol}_i}\,\text{Phos}[t] \to \frac{1}{\text{Vol}_e}\,\text{Phose}[t]\right\},$$

(*Mg-metabolite binding.*)
```
{mgatp,    Mg[t] + ATP[t] → MgATP[t]},
{mgadp,    Mg[t] + ADP[t] → MgADP[t]},
{mgb23pg,  Mg[t] + B23PG[t] → Mg$B23PG[t]},
{mgb13pg,  Mg[t] + B13PG[t] → Mg$B13PG[t]},
{mgfru16p2, Mg[t] + Fru16P2[t] → Mg$Fru16P2[t]},
{mgglc16p2, Mg[t] + Glc16P2[t] → Mg$Glc16P2[t]},
{mgphos,   Mg[t] + Phos[t] → Mg$Phos[t]},
```

(*Hb-metabolite binding.*)
```
{hbmgatp,  Hb[t] + MgATP[t] → Hb$MgATP[t]},
```

```
     {hbatp,      Hb[t] + ATP[t]  → Hb$ATP[t]},
     {hbadp,      Hb[t] + ADP[t]  → Hb$ADP[t]},
     {hbbpg,      Hb[t] + B23PG[t] → Hb$B23PG[t]},
     {hbb13pg,    Hb[t] + B13PG[t] → Hb$B13PG[t]}
   };

Save["eqns", eqns];
```

Index

A

Activators, 44
Activity coefficient, 3
Adenosine triphosphate, 105, 283
Adenylate kinase, 283
Aldolase, 267-268
Anticompetitive inhibition, 45
Arginase, 85-88
Arginine, 65-67
Argininosuccinate lyase, 84-85
Argininosuccinate synthetase, 83-84

B

Bayes' theorem, 156
Bimolecular reaction, 5-6
2,3-Bisphosphoglycerate shunt
    concentration control coefficients, 206-207
    description of, 177, 185, 194
    flux control coefficients, 204
    homeostatic strength concentrations, 213
    magnesium binding, 285-286
    rate equations, 282-283
    response coefficient of, 207-208
    response control coefficients, 212
2,3-Bisphosphoglycerate synthase-phosphatase, 282-283

C

Carbon dioxide concentration, intracellular, 292
Cell volume changes, 118-130
Cell water fraction of total cell volume, 291
Chemical kinetics, 1-2
Chi-square measure, 157
Competitive inhibition, 45
Compulsory-ordered ternary-complex mechanism, 71
ConcControlMatrix, 136, 223
Concentration control coefficients
    description of, 136-138
    erythrocyte metabolism model, 206-207
ConcResponseMatrix, 142-143, 147, 223-224
Conservation of mass
    description of, 46, 76
    erythrocyte metabolism, 187-191

Conservation relations, 110-113
ConservationRelations, 110, 112, 224
Control coefficients
    calculation of, 140-141
    concentration
        description of, 136-138
        erythrocyte metabolism model, 206-207
    description of, 134-140
    flux
        description of, 134-135
        erythrocyte metabolism model, 200-206
        numerical perturbation for calculation of, 140-141
Coupled reactions, 12-15

D
Deoxy-hemoglobin-metabolite binding, 289-290
Differential equations, parameters in systems of, 167-171
Dissociation equilibrium constant, 4
Donnan potential, 123
Donnan ratio, 291
DSolve, 9

E
Eadie-Hofstee plot, 42-43
Eigenvalues, 115-116, 191
Eisenthal and Cornish-Bowden equation, 44
Elasticity coefficients
    description of, 134, 142-145
    erythrocyte metabolism model, 211-212
Enolase, 271
Enzyme kinetics
    inhibition
        degree of, 44-45
        Michaelis-Menten equations that include effects of, 45
    mechanisms
        examples of, 59-63
        Michaelis-Menten, 45-47
        reversible Michaelis-Menten enzyme, 48-49
        steady state, 47-48, 69-75
        pH effects, 79-81
    reactions
        Eadie-Hofstee plot, 42-43
        Eisenthal and Cornish-Bowden equation, 44
        Hanes-Woolf plot, 44
        Lineweaver-Burk plot, 41-42, 50
        Michaelis-Menten equation, *See* Michaelis-Menten equation
        overview of, 39

study areas for, 53
Enzyme oligomerization, 59
Enzyme-sucrose complex, 46
EpsilonElasticityMatrix, 142-143, 224
Equilibrium constant, 4
Erythrocyte metabolism
    metabolic control analysis of
        concentration control coefficients, 206-207
        description of, 197
        elasticity coefficients, 211-212
        flux control coefficients, 200-206
        internal response coefficients, 212-213
        partitioned responses, 207-211
        response control coefficients, 207-211
            in vivo steady state, 197-198
            zero fluxes, 198-200
    model of
        conservation of mass relationships, 187-191
        description of, 175-177
        equation list, 295-297
        initial conditions for, 292-293
        parameters for, 291-292
        stoichiometry, 177-182
        time course, 182-186, 192-195
        in vivo steady state, 182-186
    schematic diagram of, 176
Euler method
    implementation of, 22-24
    improved, 28-29
    modified, 29-30
Exponential decay, 8, 12
Extracellular volume, 291

F
Figure-of-merit functions, 158
FindMinimum, 166, 168
First-order reaction, 12-15
Flux control coefficients
    description of, 134-135
    erythrocyte metabolism model, 200-206
FluxControlMatrix, 135, 224-225
FluxResponseMatrix, 142-143, 225

G
Gene, base sequence alterations of, 54
Gluconolactonase, 275
Glucose 6-phosphate, 185

Glucose 6-phosphate dehydrogenase, 274-275
Glucose phosphate isomerase, 266
Glutathione oxidation, 283
Glutathione reductase, 278-279
Glyceraldehyde phosphate dehydrogenase, 268-270
Glycolysis, 177, 202, 209

H
Half-life, 10-11
Hanes-Woolf plot, 44
Hematocrit, 291
Hemoglobin-metabolite binding, 287
Hessian method, 163
Hexokinase reaction, 180, 202, 205, 210, 265-266
Homeostatic strength, 150, 212-213

I
Inborn errors of metabolism, 54
Inhibitors, 44
Initial velocity, 40
Internal response coefficients
    description of, 149-151
    erythrocyte metabolism model, 212-213
Intracellular volume, 291
Inverse-Hessian method, 163
Inverse::luc, 201
Ionization of substrate, 79-81

J
Jacobian, 113, 115

K
K_{eq}, 54
K_m, 41

L
Lactate dehydrogenase, 272-273
    NADPH-dependent, 273-274
Lactate transport, 284
Least squares, for parameter estimation
    description of, 156-157, 170-171
    linear, 159-162
    nonlinear, 164-166
Levenburg-Marquardt algorithm, 162-164
Lifetime of reaction, 11-12
Lineweaver-Burk plot, 41-42, 50
Link matrix, 130-131

LinkMatrix, 225

M
Magnesium1,3-Bisphosphoglycerate, 286
MagnesiumFructose (1,6)-bisphosphate, 286
MagnesiumGlucose (1,6)-bisphosphate, 287
Magnesium-metabolite binding equations, 285-287
MagnesiumPhosphate, 287
Mass action, principle of, 3-4
Matrix notation, 97-102
Maximum *a posteriori*
    description of, 156-158
    nonlinear, 166-167
Membrane transport reactions, 284
Merit functions, 158
MetabolicControlAnalysis, 97-98, 103-104, 133, 230-263
Metabolic control analysis
    control coefficients
        calculation of, 140-141
        concentration, 136-138, 206-207
        description of, 134-140
        flux, 134-135, 200-206
        numerical perturbation for calculation of, 140-141
    definition of, 133
    elasticity coefficients, 134, 142-145, 211-212
    erythrocyte metabolism
        concentration control coefficients, 206-207
        description of, 197
        elasticity coefficients, 211-212
        flux control coefficients, 200-206
        internal response coefficients, 212-213
        partitioned responses, 207-211
        response control coefficients, 207-211
        in vivo steady state, 197-198
        zero fluxes, 198-200
    functions, 223-230
    matrix notation, 97
    program for, 230-262
    response coefficients
        description of, 134, 145-148
        erythrocyte metabolism, 207-211
        internal, 149-151
    summary overview of, 151-152
Metabolic pathways
    description of, 105
    interconnection patterns in, 2-3
    simulation of

cell volume changes, 118-130
conservation relations, 110-113
link matrix calculations, 130-131
matrix notation for, 97-102
steady-state concentrations, 105-110
steady state stability, 113-118
stoichiometry matrix, 102-105
tasks involved in, 155
time-dependent behavior of multi-enzyme systems, 95-97
urea cycle, 81-89
Metabolism
erythrocyte, *See* Erythrocyte metabolism
inborn errors of, 54
Michaelis constant, 40
Michaelis-Menten equation
definition of, 39-40
enzyme inhibition effects included in, 45
enzyme mechanisms, 45-47
formula for, 40
K_m, 41
nonlinear regression, 165-166
plotting of, 165
reversible, 48-49
time courses solved using, 61
V_{max}, 41
Michaelis-Menten reaction
pre-steady-state, 57-59
progress curve of, 55
Mixed inhibition, 45
MMatrix, 114, 226
Molecularity, 5
Monte Carlo simulation, 172
mRNA, 2
Multi-enzyme systems, time-dependent behavior of, 95-97
Multiple-coupled reaction, 16-18
Multiple equilibria, 75-79

N
NADPH-dependent lactate dehydrogenase, 273-274
NDSolve, 35, 55-56
NDSolveMatrix, 100-101, 107, 226
Nernst equation, 123
NMatrix, 103, 226
Non-glycolytic NADH consumption, 283
Nonlinear, 18
Nonlinear least squares, 164-166
NSteadyState, 109, 188, 226-227

Nuclear magnetic resonance spectrometer, 39, 53
Numerical integration
    definition of, 18
    Euler method
        implementation of, 22-24
        improved, 28-29
        modified, 29-30
    general, 20-21
    multi-step methods, 24
    NDSolve, 35
    one-step methods, 26
    overview of, 22-25
    predictor-corrector methods
        corrector, 33-34
        description of, 24-25, 31-32
        h value, 34
        predictor, 32
    Runge-Kutta methods
        general, 30-31
        higher-order, 31
        overview of, 26-28
    single-step methods, 24
    stiffness, 35
    Taylor series solution, 25-26, 28-29
Numerical perturbation method, for control coefficient calculations, 140-141

O
Order of a reaction, 5-6
Order of the method, 26
Ornithine carbamoyl transferase, 83
Oxy-hemoglobin-metabolite binding, 287-288

P
Parameter estimation
    approaches to, 155-156
    differential equations, 167-171
    goal of, 155-156
    least squares
        description of, 156-157, 170-171
        linear, 159-162
        nonlinear, 164-166
    maximum *a posteriori*
        description of, 156-158
        nonlinear, 166-167
        overview of, 155
    parameters
        optimal, 171

variances of, 172-173
rate equations
    description of, 159
    Levenburg-Marquardt algorithm, 162-164
    linear least squares, 159-162
    nonlinear estimation, 162-164
PartialConcResponse, 148, 227
PartialFluxResponse, 148, 227
PartialInternalConcResponse, 148, 228
PartialInternalFluxResponse, 148, 228
Pentose phosphate pathway
    description of, 177, 204
    rate equations, 274-282
pH
    elasticity coefficients, 211
    enzyme kinetic parameters affected by, 79-81
    intracellular, 292
Phosphate transport, 284
Phosphofructokinase, 266-267
6-Phosphogluconate dehydrogenase, 276-277
6-Phosphogluconolactone hydrolysis, 275-276
Phosphoglycerate kinase, 270
Phosphoglycerate mutase, 271
PiElasticityMatrix, 142, 228-229
Potassium concentrations, intracellular, 292
Predictor-corrector methods
    corrector, 33-34
    description of, 24-25, 31-32
    h value, 34
    predictor, 32
Primary reactant, 4
Principle of mass action, 3-4
Prior probability, 156
Progress curves, of Michaelis-Menten reaction, 55
Pure noncompetitive inhibition, 45
Pyruvate kinase, 272
Pyruvate transport, 284

R
Rate constant
    definition of, 3
    second-order, 63-64
    units of, 6
Rate equation deriver, 217-222
RateEquation function, 67-68
Rate equations
    erythrocyte metabolism, 265-290

parameter estimation
    description of, 159
    Levenburg-Marquardt algorithm, 162-164
    linear least squares, 159-162
    nonlinear estimation, 162-164
steady-state
    deriving of, 65-67
    enzymic reactions modeled using, 180
    examples of, 66-67
    RateEquation function, 67-68
Reaction, *See also specific reaction*
    bimolecular, 5-6
    coupled, 12-15
    extent of, 6-7
    first-order, 12-15
    half-life of, 10-11
    lifetime of, 11-12
    multiple-coupled, 16-18
    non-first-order, 18-20
    time courses of, 7-20
Reaction order, 5-6
Reaction rate, 63-64
Rectangular hyperbola, 46-47, 50
Regress[*data*, {1, *x*,} {*x*}], 159
Regulatory enzymes, 50-51
Response coefficients
    description of, 134, 145-148
    erythrocyte metabolism model, 207-211
    internal
        description of, 149-151
        erythrocyte metabolism model, 212-213
Reversible Michaelis-Menten enzyme, 48-49
Ribulose-5-phosphate epimerase, 279-280
Ribulose-5-phosphate isomerase, 280
Runge-Kutta methods
    general, 30-31
    higher-order, 31
    overview of, 26-28
    second-order, 31

S
Saturation, 46
Second-order rate constant, 63-64
Sensitivity analysis, 133
SMatrix, 103, 229
Spectrophotometer, 39
Stability, 114, 229

Statistics `LinearRegression, 159
Statistics `NonlinearFit NonlinearRegress[*data*, {1, *x*,} {*x*²}], 164
SteadyState, 106, 109
Steady state
    concentration determinations, 105-110
    enzyme kinetics, 47-48
    erythrocyte metabolism model, 191
    hexokinase at, 205
    parameters, deriving expressions for, 69-75
    stability of, 113-118
    in vivo, for erythrocyte metabolism, 182-186, 197-198
Steady-state rate equations
    deriving of, 65-67
    enzymic reactions modeled using, 180
    examples of, 66-67
    RateEquation function, 67-68
Stiffness, 35
Stoichiometry
    erythrocyte metabolism, 177-182
    matrix, 102-105
    vectors for, 98-99
StoichiometryMatrix, 229-230
Substrate, ionization of, 79-81
Summation theorem, 136-137

T
Taylor series, 25-26, 28-29
Time course
    erythrocyte metabolism model, 182-186, 192-195
    intracellular calcium ion, 128
    intracellular potassium ion, 128
    Michaelis-Menten equation, 61
    of reactions, 7-20
Transaldolase, 281-282
Transcription factors, 2
Transketolase, 280-281
Triosphosphate isomerase, 268
Turnover number, 47, 59, 70

U
Uncompetitive inhibition, 45
Unitary rate constants
    calculating of, 69
    definition of, 3, 6
    values for, upper limits of, 63-65
Urea, 66
Urea cycle, 81-89

V
VMatrix, 103, 230
V_{max}, 41, 70

Z
Zero fluxes, 198-200